Big Data Conference (BDC 2017)

AA006281

Journal of Physics: Conference Series Volume 913

Moscow, Russia
15 September 2017

ISBN: 978-1-5108-5040-8
ISSN: 1742-6588

Printed from e-media with permission by:

Curran Associates, Inc.
57 Morehouse Lane
Red Hook, NY 12571

Some format issues inherent in the e-media version may also appear in this print version.

Printed by Curran Associates, Inc. (2017)

For permission requests, please contact the Institute of Physics at the address below.

Institute of Physics
Dirac House, Temple Back
Bristol BS1 6BE UK

Phone: 44 1 17 929 7481
Fax: 44 1 17 920 0979

techtracking@iop.org

Additional copies of this publication are available from:

Curran Associates, Inc.
57 Morehouse Lane
Red Hook, NY 12571 USA
Phone: 845-758-0400
Fax: 845-758-2633
Email: curran@proceedings.com
Web: www.proceedings.com

Table of contents

Volume 913

BigData Conference (Formerly International Conference on Big Data and Its Applications)

15 September 2017, Moscow, Russian Federation

Accepted papers received: 16 October 2017
Published online: 25 October 2017

Machine learning and artificial intelligence

Distributed systems

IOP Conf. Series: Journal of Physics: Conf. Series **913** (2017) 011001 doi:10.1088/1742-6596/913/1/011001

Preface

The 2017 4rd Big Data Conference (BDC 2017, formerly International Conference on Big Data and Its Applications) was successfully held on September 15, 2017 in Moscow, Russia. The conference provided a leading forum for disseminating the latest results in Big Data Research, Development, and Applications from business, technological and scientific points of view. For the scientific track of the conference solicit high-quality original research papers (and significant work-in-progress papers) in any aspect of Big Data with emphasis on 5Vs (Volume, Velocity, Variety, Value and Veracity), including the Big Data challenges in scientific and engineering, social, sensor/IoT/IoE, and multimedia (audio, video, image, etc.) big data systems and applications. There were 3 major sections of the scientific track of the conference:

- Fundamentals of Data Science
- Big Data Infrastructure
- Big Data Security, Privacy and Trust
- Big Data Management
- Big Data Search and Mining
- Big Data Applications

This volume of Journal of Physics: Conference Series is a compilation of the accepted papers in BDC 2017 and represents contributions that were presented in the scientific track of the conference.

On behalf of the organization committee I would like to thank all of the conference sponsors, partners and volunteers who made conference possible.

Looking forward to meet you during BDC 2018.

On behalf of organizing and program committees of the BDC 2017

Igor Balk, BDC 2017 co-chair.

IOP Conf. Series: Journal of Physics: Conf. Series **913** (2017) 011002 doi:10.1088/1742-6596/913/1/011002

Peer review statement

All papers published in this volume of *Journal of Physics: Conference Series* have been peer reviewed through processes administered by the proceedings Editors. Reviews were conducted by expert referees to the professional and scientific standards expected of a proceedings journal published by IOP Publishing.

IOP Conf. Series: Journal of Physics: Conf. Series **913** (2017) 012001 doi:10.1088/1742-6596/913/1/012001

AI User Support System for SAP ERP

Vladimir Vlasov[1], Victoria Chebotareva[1], Marat Rakhimov[1] and Sergey Kruglikov[2]

[1]JSC Sberbank, 620000, Ekaterinburg, Kuibysheva 67, Russia
[2]Ural Federal University, Ekaterinburg, Russia

e-mail: vevlasov@sberbank.ru, svk@imm.uran.ru

Abstract. An intelligent system for SAP ERP user support is proposed in this paper. It enables automatic replies on users' requests for support, saving time for problem analysis and resolution and improving responsiveness for end users. The system is based on an ensemble of machine learning algorithms of multiclass text classification, providing efficient question understanding, and a special framework for evidence retrieval, providing the best answer derivation.
Keywords: Multi-class Text Classification, Natural Language Processing, Question Answering, Automated User Support, SAP ERP.

1. Problem Description

Enterprise resource planning (ERP) is business process management software that allows organization to use a system of integrated applications to manage the business and automate many functions, related to technology, services and human resources.

For a large scale business running complex IT system such as SAP ERP, it is typical to have hundreds of thousands of users' requests annually reported to the help desk verbally by telephone or via specialized help desk systems, such as Service Manager (SM) or Service Desk.

There are usually two types of problems that are reported: errors, arising from incorrect users' actions or system failures and requests for information, advice on a particular user case. Both of these types of problems interrupt a normal course of operations requiring 1–2 days for a problem being solved, leading to additional costs on support team maintenance.

Users' requests are messages in natural language, containing some kind of information about the problem which is relevant in the users opinion. In fact, types of problem descriptions range from a deep problem analysis conducted by a professional user to less informative and sometimes useless or misleading messages.

The ability to understand human posed questions and to provide relevant answers refers to one of the most challenging AI problems – Question Answering (QA). QA is a computer science discipline within the fields of information retrieval and natural language processing. In comparison with the open-domain QA systems, such as IBM Watson, designed to challenge human champion in the Jeopardy! Game [1], dealing with various problems of producing sensible answers out of general knowledge: from lexical answer type determination to massively parallel hypothesis scoring, closed-domain QA systems deal with specific knowledge which can be formalized in ontologies.

IOP Conf. Series: Journal of Physics: Conf. Series **913** (2017) 012001 doi:10.1088/1742-6596/913/1/012001

The problem of implementing machine learning algorithms to help desk automation is described in [2].

Thus building an intelligent user support system is a closed-domain QA problem. In this paper we present a description of such QA system with application to a particular domain of SAP ERP user support, based on machine learning algorithms of text classification.

2. Intelligent user support system architecture

The high-level architecture of the intelligent user support system is presented in figure 1.

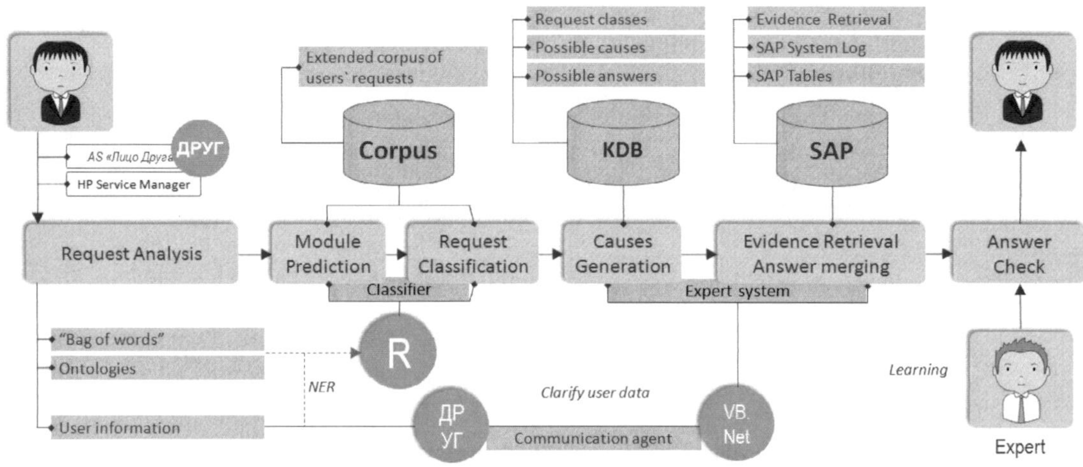

Figure 1. The high-level architecture of the intelligent user support system for SAP ERP

The main steps in a high-level architecture of the proposed intelligent user support system.

1. Content acquisition including creation, cleaning and extending of a corpus of users' requests.
2. Request analysis including SAP module determination and deep request classification.
3. Knowledge database development in order to map classes of requests on sets of possible causes.
4. Generation of a set of possible causes for the particular type of problem.
5. Gathering evidence from SAP to evaluate hypothesis about candidate causes of problem.
6. Answer merging w.r.t. available information from SAP tables, system log and user information.
7. Answer checked by an expert. If:
   a. Correct, then the answer sent to user.
   b. Incorrect, then the request and the correct class* added to the training set in R. An algorithm relearned and updated.

3. Request Classi_cation Algorithm

A classification of requests is a standard text classification task performed by various machine learning algorithms providing multi-class classification [3–6].There were attempts to implement more sophisticated algorithms for this task such as recurrent convolutional neural network [7], but the increase in accuracy was not significant.

Raw data comes from SAP ERP support database of one large Russian company. All requests are primarily divided into six main categories (modules) in accordance with business processes realized in SAP ERP environment. Requests out of each module have been labeled by human experts to one of the classes from the set of classes (C).

Main categories (modules) are the following.

1. Purchases and agreements.

* New classes of requests potentially can arise.

2. Property and material management.
3. Budget planning and execution.
4. Real estate management and construction.
5. Business trips and hospitality expenses and Payments.
6. User Access management.

A class label is a short text of typical SAP ERP user problem: an error message or general user request problem type. Each class is constructed in a manner, that requests attributed to one class, share the same set of possible causes and corresponding answers.

Some class labels examples in the Purchases and Agreements category: "How to execute the request for supply", "Error: applicant is unavailable", "Specification not displayed in agreement", "no resource found when creating a request for supply", etc.

Users can express one problem in different ways, for example, by the following examples. Consider a user is faced with a problem "no resource found when creating a request for supply". Then request can be formulated as: "when creating Request for Supply I cannot select the required resource" or "when creating Request for Supply the program does not find the resource"

The number of classes by categories is: $l_1 = 81$, $l_2 = 43$, $l_3 = 26$, $l_4 = 23$, $l_5 = 33$, $l_6 = 9$. Total number of classes is $l = 215$. Each request can be related only to one class. The number of requests in the labeled sample is the following: $m_1 = 4168$, $m_2 = 2824$, $m_3 = 1698$, $m_4 = 867$, $m_5 = 1609$, $m_6 = 1388$. Total number of requests in a labeled sample $m = 12554$.

The key aspect to an efficient classification algorithm is the preparation of the text corpus [2]. Corpus – is a complete collection of texts of user requests, containing raw texts with class labels. Cleaning and extension of a corpus include a sequence of procedures, developed in R via different packages of text mining ("tm"), stemming ("SnowballC"), splitting, combining data ("plyr"):

1. Conversion to lowercase.
2. Deletion of doubled whitespaces, numbers, punctuation.
3. Deletion of stopwords ("and", "or", "good morning", "thank", "please").
4. Reducing words to their word stems, base form (stemming).
5. N-gram retrieval (n-gram – is a contiguous sequence of n items from a given sequence of text), ("create new request for supply", "insufficient budget").
6. Unification of synonymic constructions

Building a term-document matrix (TDM), containing each request as a vector of numerical attributes, corresponding to tokens encountered inside requests.

In the course of the work, it was concluded that not all stopwords should be removed. The word "not/no" turned out to be meaningful within the framework of our task. And for module "User Access management" prepositions proved to be important. Without them you cannot correctly determine the class.

To reduce the number of words in corpus and to improve accuracy of classification you can combine words into artificial terms. You can combine words in accordance with one of three principles: acronym expansions: "RFS" – "request for supply", synonyms in the sense of the Russian language: "storekeeper" – "warehouse manager" and synonymous words in the context of SAP: "budget indicator red" – "insufficient budget". Such synonyms where chosen manually. Users can formulate one problem in different ways, sometimes with the most unusual words.

A TDM is a sparse matrix, i.e. most of its elements are zeros. To prevent overfitting and ensure classes generalization and interpretability, we should remove sparse terms for each class. Selection of a appropriate sparsity threshold parameter is a subject to optimization. This value of a threshold delivering the best average accuracy is chosen for each class. We use sparsity thresholds 60% for all classes. Also we use TF-IDF to assess the importance of words in requests. But this method did not give a good result in our problem: specific words from small classes remained invaluable. To eliminate this imbalance, the TF-SLF was applied. This method is based on the fact, that the term is important within a category if it occurs in most documents of this category. This approach allows

IOP Conf. Series: Journal of Physics: Conf. Series **913** (2017) 012001 doi:10.1088/1742-6596/913/1/012001

you to estimate the weight of keywords for all classes and reduce the weight of words that are key for several classes.

We train each of the following algorithms on a training set and test it on a test set randomly sampled from each class in proportion of 75/25 %.

1. Naïve Bayes – NB
2. k Nearest Neighbors – kNN
3. Support Vector Machine – SVM
4. Multinominal Logistic Regression or Maximum Entropy – MaxEnt.

A brief description of each algorithm used for classification is presented in the following section.

3.1. Naïve Bayes Algorithm (NB)

NB – is one of the simplest yet effective method of multi-class classification [5] based on Bayesian Rule. The probability of class *c* given document *d* is determined by the following equation (1):

$$p(c|d) = \frac{p(d|c) \cdot p(c)}{p(d)} \tag{1}$$

Thus, a class of request is determined by (2):

$$c' = \underset{c \in C}{\operatorname{argmax}}(p(c|d)) = \underset{c \in C}{\operatorname{argmax}}\big(p(d|c) \cdot p(c)\big) = \underset{c \in C}{\operatorname{argmax}}\big(p(x_1, x_2, \dots, x_n|c) \cdot p(c)\big) \tag{2}$$

where $x_1, x_2, \dots, x_n$ – is a vector representation of *d* (“bag of words” representation).

Making an assumption of feature x_i conditional independence, which is obviously wrong (terms «budget» and «control» more often appear together), but allows joint probability $p(x_1, x_2, \dots, x_n|c)$ to be represented as a product of probabilities $p(x_i|c)$. Then formula (2) takes its final representation (3):

$$c_{NB} = \underset{c \in C}{\operatorname{argmax}}\, p(c_j) \prod_{x \in X} p(x|c) \tag{3}$$

The probabilities in (3) are substituted by their frequency-based estimators: $P(c_j)$ is estimated as a ratio of requests related to class c_j to total number of requests, $p(x|c)$ is estimated as a fraction of times term *x* appears among all terms in requests of class c_j.

Classification made by NB is a good benchmark for other more sophisticated algorithms.

3.2. K Nearest Neighbors (KNN)

KNN – is an algorithm that returns the class which contains the most similar training request to the one that is being classified. The use of different similarity functions forms different types of kNN. We use the standard Euclidean distance (4).

$$\rho(u, x_i) = \left(\sum_{j=1}^{n} \left| u^j - x_i^j \right|^2 \right)^{1/2} \tag{4}$$

where x_i^j – *i*-th neighbor of the object *u*.

For each *u*, its neighbors from the training sample can be arranged in ascending order w.r.t. ρ. If $c_u^{(i)}$ – is the class of *i*-th neighbor of object *u*, then the class of u is determined by the equation (5):

$$c_{kNN} = \underset{c \in C}{\operatorname{argmax}} \sum_{i=1}^{m} [c_u^{(i)} = c] \cdot w(i, u) \tag{5}$$

Choosing $w(i, u)$ one can form various types of kNN algorithm: exponentially weighted NN, Parzen window kNN, etc. [8]. The number *k* of NN is modelled by leave-one-out cross-validation.

IOP Conf. Series: Journal of Physics: Conf. Series **913** (2017) 012001 doi:10.1088/1742-6596/913/1/012001

3.3. Support Vector Machine (SVM)

SVM – is the non-probabilistic classification method introduced by V. Vapnik, which constructs a hyperplane in a high-dimensional feature space by empirical risk minimization [8]. SVM – is a binary classification algorithm. Traditional way of performing multiclass classification with SVM is combining several binary "one-against-all" or "one-against-one" SVM classifiers [5]. We have k (by the number of classes) similar "one-against-all" optimization tasks (6):

$$\min_{w,b,\xi} \frac{1}{2} w^T w + C \sum_{i=1}^{l} \xi_i$$
$$w^T \phi(x_i) + b \geq 1 - \xi_i, \text{if } y_i = m, \quad (6)$$
$$w^T \phi(x_i) + b \leq -1 + \xi_i, \text{if } y_i \neq m,$$
$$\xi_i \geq 0$$

where y_i– class of x_i, ϕ – kernel function. C – penalty parameter.

Thus a class of request is derived by (7):

$$c_{SVM} = \underset{c \in C}{\operatorname{argmax}}((w^c)^T \phi(x) + b^c) \quad (7)$$

As long as in the text classification problem classes are usually linearly separable, SVM with linear kernel should perform well for text categorization [5].

3.4. Maximum Entropy (MaxEnt)

Multinominal logistic regression or maximum entropy [9] – probabilistic log-linear classifier, maximizing log-likelihood of a set of weighted features from input data. The class c of the request x is determined as maximum probability $p(c|x)$, that is computed by (8)

$$p(c|x) = \frac{\exp(\sum_{i=1}^{N} w_i f_i(c,x))}{\sum_{c' \in C} \exp(\sum_{i=1}^{N} w_i f_i(c',x))} \quad (8)$$

where $f_i(c,x)$ – i-th binary feature of request x given class c. This feature takes the value of 1, if a term is present in a request and the class is c, and zero otherwise.

Special regularization methods are also used to fix the problem of overfitting. In our work we used the implementation of MaxEnt algorithm in R written by T. P. Jurka [6, 10].

The results in terms of accuracy and its variation on local tests are presented in Table 1. The results for parameters with the best performance on the test set are reported.

Table 1. Mean and Variation of accuracy on local tests (%)

Support group (SAP module / process)	NB	kNN	SVM	MaxEnt
Purchases and agreements	76.4 (2.1)	85.4 (1.4)	**87.8 (1.9)**	87.1 (2.1)
Property and material management	76.6 (2.5)	86.6 (1.6)	89.5 (2.1)	**90.3 (1.5)**
Budget planning and execution	64.6 (3.4)	77.4 (3.5)	85.6 (1.5)	**87.3 (2.3)**
Real estate management and construction	88.8 (2.9)	91.1 (3.2)	**93.5 (2.9)**	92.4 (2.9)
Business trips and hospitality expenses and Payments	88.3 (1.9)	92.2 (2.1)	**94.5 (1.8)**	93.9 (1.4)
User Access management	82.1 (2.6)	91.9 (2.98)	**95.1 (2.6)**	56.3 (3.6)

IOP Conf. Series: Journal of Physics: Conf. Series **913** (2017) 012001 doi:10.1088/1742-6596/913/1/012001

As it is shown in table 1, there is no single choice for the type of the classification algorithm. MaxEnt and SVM are the most popular choice. So by far we completed the steps that allowed the proposed system to understand what kind of problem a user was trying to express in his request.

4. Answering the request and error analysis

After defining the problem class the system gives the expert a prepared answer for this class. The answer for class can be unambiguous (for the request "how to get the role of a material responsible person") or may consist of several recommendations depending on the cause of the problem (for the request "no resource found when creating a request for supply"). In this later case it is necessary to conduct additional analysis. The analysis consists of hypotheses testing for the candidate causes of the problem. Hypotheses testing imply check-listing a set of rigid rules predefined by the expert in order to reduce the number of potential causes and the corresponding answers.

For each class of requests there are certain parameters to retrieve from SAP ERP system and to compare with the values from the user request. In case of missing user parameters in the initial request, the system either looks them up in a system log or asks user to send information.

After the evidence has been retrieved the process of "pruning the trees" of candidate causes is performed and a single answer (perhaps combined of several reasons and recommendations) is given. The answer is then verified by a human expert and in case it is incorrect – reclassified by an expert.

Thus the process of forming a response can be represented as follows:

- Collecting necessary data to test hypotheses about the cause of the problem (from a user request).
- Requesting SAP database for necessary parameters specified for each class.
- Verification of criteria.
- Rejection of incorrect hypotheses.
- Formulation of the final answer.

For example we will take the class from the module "Purchases and agreements": "no resource found when creating a request for supply". The data required for the hypothesis testing are the number of the resource and the number of request for supply. If these data are present in the message, we can proceed to step 2. In case of the absence of necessary information, the system will send a request to the user or the system looks it up in a system log. When the number of request for supply and resource are obtained, we proceed to testing the hypotheses. In this case there are 3 of them.

1. Expiration date of the resource > date of consumption in the request for supply.
2. Invalid resource number.
3. A resource cannot be used in the request for supply.

With respect to a certain request ID, for each of hypotheses the system retrieves specific parameters (Expiration date of the resource, number of the resource, number of the request for supply) and in case the conditions of the hypothesis are matched, it is regarded as true.

If this hypothesis is confirmed, then the user receives a particular response, such as: "Good day! Select another resource".

As for incorrectly classified appeals, they are analyzed. The training set and the classification algorithm is updated on weekly basis with regard to the error analysis. The most common reasons of errors in classification of requests are the following:

- Uncommon synonym for the specific term was used in the request. In this case this synonym is added to a term definition.
- Presence of keywords from different classes in one request. Usually this is a problem of long requests, when the user formulates several questions in one request. Those requests are treated individually by expert.

IOP Conf. Series: Journal of Physics: Conf. Series **913** (2017) 012001 doi:10.1088/1742-6596/913/1/012001

5. Conclusions

The paper introduces a machine-powered user support system for SAP ERP. It describes the main steps of the process of corpus preparation, problem understanding and answer derivation combined in a simple architecture, providing the system to evolve.

The system is scalable on various problems of closed-domain question answering in different spheres and requires joint efforts of human experts for data preparation, building a knowledge database and text mining techniques and machine learning algorithms of multi-class classification to potentially reach a high-level accuracy on understanding and answering natural language requests.

References

[1] Ferrucci D, Brown E et al 2010 Building Watson: An Overview of the DeepQA Project *AI Magazine* **31 (3)**
[2] Logan D and Kenyon J 1992 HelpDesk: Using AI to Improve Customer Service *In Proceedings AAAI-92* p 37–53
[3] D'Auria T *KNN Text Classification in R* (Video Tutorial)
[4] Manning C, Raghavan P and Schütze H 2008 *Introduction to Information Retrieval* (Cambridge University Press)
[5] Joachims T 1998 Text categorization with support vector machines: Learning with many relevant features *In Proceedings of the European Conference on Machine Learning* (Springer)
[6] Nigam K, Lafferty J and McCallum A 1999 Using maximum entropy for text classification *IJCAI-99Workshop on Machine Learning for Information Filtering* p 61–67
[7] Lai S, Xu L, Liu K and Zhao J 2015 Recurrent Convolutional Neural Networks for Text Classification *In Proceedings of the Twenty-Ninth AAAI Conference on Artificial Intelligence* **333** p 2267–2273
[8] Cortes C and Vapnik V 1995 Support-vector network *Machine Learning* **20** p 273–297
[9] Jurafsky D and Martin J H 2009 *Speech and Language Processing: An Introduction to Natural Language Processing, Speech Recognition, and Computational Linguistics* (2nd edition, Prentice-Hall)
[10] Jurka T P 2012 Maxent: An R Package for Low-memory Multinomial Logistic Regression with Support for Semi-automated Text Classification *The R Journal* **4(1)** p 56–59

IOP Conf. Series: Journal of Physics: Conf. Series **913** (2017) 012002 doi:10.1088/1742-6596/913/1/012002

A multidimensional approach to assessing the elemental status of an organism

S Akimov, P Vedeneev, E Kiyaeva, I Laryushina, S Notova and A Pishchukhin

Orenburg State University, Russia, Orenburg, Pobedi avenue, 13

e-mail: Pishchukhin55@mail.ru

Abstract. Multidimensional space is a convenient means of representing large amounts of information. This fully applies to information on the elemental status of population groups. The novelty of the approach of this study is based on the fact that the totality of the weight parts of all elements of the periodic table together makes up the weight of a person. In a multidimensional space, organisms with the same weight have the same sum of coordinates and are located on one hyperplane. Since for any norms it is important to have a ratio between the quantities that reflect the content of the chemical elements – a ray becomes the standard, for each point of which the ratio between the coordinates is observed. Large amounts of data will adequately represent the proximity an organism to one or another class, that is, increasing the accuracy of diagnosing the elemental status. The algorithm for diagnosing, therefore, should include finding the corresponding hyperplane, the point of intersection with the ray and determining the proximity to this point.

1. Introduction

With the advent of modern technology, diagnostic possibilities extend greatly. Volumes of the analyzed information grow, including the data of functional, biochemical, genetic and molecular studies, which greatly complicates the work of the doctor. Using of multidimensional space as a way to represent a large amount of heterogeneous information is the most rational. In the multidimensional space, the target area of the "healthy organism" and the current state of the subject are determined. This allows you to compare the methods in terms of their effectiveness and helps to purposefully formulate a strategy of recovery treatment for a particular patient. The application of this method makes it possible to take into account the adaptation processes in assessing the health status of the individual, including relying on the accumulated data on the elemental status – one of the fundamental indicators of health [1–3]. For this purpose, in the multidimensional space, as target, it is possible to identify local areas reflecting the adaptation of the indigenous population to the conditions of a particular biogeochemical province.

The study is devoted to the development of a multidimensional model of space in which it is possible to identify regional norms of elemental status for different age and gender categories of the population. Now there is a problem of diagnostic evaluation and complex interpretation of a huge number of health indicators. In this regard, there is pressing need for data systematization and automation of the processing of data, and development of a computer support system for making medical decisions. A large amount of objective information, on the one hand, allows to rationalize the diagnostic process and formulate a treatment strategy, and on the other hand, the absence of a systematizing model makes a complete analysis of the obtained research results impossible.

IOP Conf. Series: Journal of Physics: Conf. Series **913** (2017) 012002 doi:10.1088/1742-6596/913/1/012002

This contradiction allows to formulate a problem of imperfection of the theoretical foundations for the development of tools for diagnosing the elemental status of the organism It is necessary to create a model for multidimensional and systematic representation of available information using methods of multidimensional geometry.

2. Theory

Multidimensional geometry is associated with pattern recognition and with automatic classification almost from the moment of its origin. The method of the multidimensional feature space used in the classification is one of the most important in the theory of pattern recognition in our time. It is most often associated with the decision of the question of the proximity of a given object displayed in a multidimensional space with a point with coordinates equal to the values of the characteristics for the so-called standard – the center of the cloud of points reflecting the objects of the given class and grouped by virtue of a common property or finding a given point in a given region.

Ideas of multidimensional space were expressed in the 18th century by I. Kant and J. D'Alembert, and multidimensional geometry was constructed by A. Cayley, G. Grassmann, and L. Schlefli in the before last century [4]. The advantage of the method with small dimensions is visualization. Even three-dimensional space can be illustrated in two-dimensional paintings, drawings and figures, which greatly helps the study. However, with more measurements, the visibility of this method is lost. Nevertheless, the multidimensional representation [5], supplemented for example by the agent-oriented approach [6], in which the patient, the method of treatment and the physician play an active role, have great prospects.

In this paper we consider a problem in which objects belonging to a given class have the same relationship between coordinates, and as the general property the sum of all coordinates is chosen. This formulation follows, for example, from the practical task of determining the elemental status of a person, in which the weight components of all elements of the periodic table are plotted, then the sum of all the coordinates of a single point is equal to the weight of a person.

Investigating this question, we note that points with the same sum of coordinates are located on the plane described by equation

$$\sum_{i=1}^{n} x_i + D = 0 \tag{1}$$

A distinctive property of this formula is the unit coefficients for all coordinates.

For two coordinates, this property is possessed by a straight line that runs at an angle of 45 degrees to the coordinate axes. The figure shows that for each point of this line, one coordinate increase results in an equal reduction of the other, so that the sum will be saved.

Similar arguments in three-dimensional space lead to a plane inclined at identical angles to all coordinate planes.

On the other hand, the locus of points having the same relationship between coordinates is a ray emanating from the origin, described by equation

$$x_i = t x_{Ai} \tag{2}$$

where x_{Ai} are the components of the radius of the vector specifying the line, t is an arbitrary parameter that allows one to run through this ray.

In order to classify a given point B, that is, to answer the question whether the given point belongs to the class given by the ray (2), it is necessary first to construct a plane with the same sum of coordinates as the point B. To do this, we substitute its coordinates in equation (1) and we find the parameter D. Finally, the equation of the plane is

$$\sum_{i=1}^{n} x_i - \sum_{i=1}^{n} x_{Bi} = 0 \tag{3}$$

IOP Conf. Series: Journal of Physics: Conf. Series **913** (2017) 012002 doi:10.1088/1742-6596/913/1/012002

Now it is necessary to find the point C of intersection of the given ray with the plane defined in this way, for which we substitute the coordinates (2) in equation (3) and determine the required point C in terms of the parameter t

$$t = \frac{\sum_{i=1}^{n} x_{Bi}}{\sum_{i=1}^{n} x_{Ai}} \qquad (4)$$

Now it seems that you can simply enter the degree of proximity of a given point to the point of intersection (or the point's entry into the circle of a given radius with the center at the intersection point, see the picture) as a criterion for belonging to a given class and thereby solve the question of its belonging to a given class. However, this method works only when the values of all the coordinates are close to each other. In the case when there is at least one coordinate with a small absolute value, the circle in the figure ceases to be a circle.

It would seem that this problem can be solved by scaling, but this breaks the constraint (1), in which the coefficients at all coordinates must be unitary. The only way out of the situation that emerges can be a coordinate-wise criterion for assigning a given point to a selected class.

In this case we take the coordinates of the intersection point and for each of them we set the degree of proximity, the latter should depend on the absolute value of the coordinate, for example, in a linear fashion. The slide shows that the range of occurrence is greater, the larger the absolute value of the coordinate is.

The criterion for the occurrence of a point in a given class will take the form of a system n of non-strict inequalities of the type

$$|x_{Bi} - x_{Ci}| \leq \alpha x_{Ci}, \text{ with } i = 1, \ldots, n \qquad (5)$$

The question of the occurrence of a point in a given class is solved positively only when all these non-strict inequalities are satisfied.

3. Method

We will especially discuss the problem of determining the standard of the class in view of the fact that it is not a point, as in most often happens, but a ray. Objects belonging to a given class usually form a cloud of points and there are two possible ways of constructing the desired ray.

The first is based on this or that averaging and carrying out the ray from the origin to the resulting averaged point.

$$x_{Ai} = \frac{\sum_{i=1}^{n} x_i}{n} \qquad (6)$$

The second implies carrying out a ray through a cloud of points using the least squares method, that is, so that the sum of the squares of distances from points to the ray is minimal.

Let the ray be described by equation (2). Then for each j-point from the cloud it is necessary to find the shortest distance to this ray. First, we construct a plane passing through this point perpendicular to the given ray

$$\sum_{i=1}^{n} x_{Ai}\, x_i - \sum_{i=1}^{n} x_{Ak} x_{kj} = 0 \qquad (7)$$

We find the intersection point of the ray with the plane, given by the value of the parameter t from equation (2)

$$t = \frac{\sum_{i=1}^{n} x_{Ai} x_{ij}}{\sum_{i=1}^{n} {x_{Ai}}^2} \qquad (8)$$

Now you can find the distance between the j-point and the intersection point

$$d = \sqrt{\sum_{i=1}^{n} \left(x_{ij} - \frac{\sum_{i=1}^{n} x_{Ai} x_{ij}}{\sum_{i=1}^{n} {x_{Ai}}^2} x_{Ai}\right)^2} \qquad (9)$$

Then the sum of the squares of the distances to all m points of the cloud will be

$$F(x_{Ai}) = \sum_{j=1}^{m}\sum_{i=1}^{n}(x_{ij} - \frac{\sum_{i=1}^{n} x_{Ai}x_{ij}}{\sum_{i=1}^{n} x_{Ai}^2} x_{Ai})^2 \quad (10)$$

Differentiating this sum over each of the components of the vector forming the ray and equating the derivative to zero, we obtain n equations for the determination of all components.

$$\frac{\partial F(x_{Ai})}{\partial x_{Ai}} = \sum_{j=1}^{m}\sum_{i=1}^{n} 2[x_{ij} - \frac{\sum_{i=1}^{n} x_{Ai}x_{ij}}{\sum_{i=1}^{n} x_{Ai}^2} x_{Ai}] \times \frac{2(\sum_{i=1}^{n} x_{Ai}x_{ij})x_{Ai}^2 - (x_{ij} + \sum_{i=1}^{n} x_{Ai}x_{ij})\sum_{i=1}^{n} x_{Ai}^2}{(\sum_{i=1}^{n} x_{Ai}^2)^2} = 0 \quad (11)$$

To make zero, this expression can be either expressed in square brackets, or the numerator of a fraction. In the first case it is necessary to solve a system of n square equations

$$x_{ij}x_{Ai}^2 + (\sum_{k=1,k\neq i}^{n} x_{Ak}\, x_{kj})x_{Ai} - x_{ij} = 0, i = 1, \dots, n \quad (12)$$

In the second case we arrive at a system of cubic equations

$$x_{ij}x_{Ai}^3 + (\sum_{k=1,k\neq i}^{n} x_{Ak}\, x_{kj} - x_{ij})x_{Ai}^2 - (\sum_{k=1,k\neq i}^{n} x_{Ai}^2)x_{ij}x_{Ai} - x_{ij}\sum_{k=1,k\neq i}^{n} x_{Ai}^2 - (\sum_{k=1,k\neq i}^{n} x_{Ai}^2)(\sum_{k=1,k\neq i}^{n} x_{Ak}\, x_{kj}) = 0, i = 1, \dots, n \quad (13)$$

Obviously, these equations can be solved numerically for large dimensions.

A specific task should suggest which of these ways to choose for practical use. For the task formulated at the beginning of the report, the second method is more suitable, in spite of its cumbersome nature. This is due to the heterogeneity of the statistical material with respect to the weight of the subjects.

4. Data and modeling results

The mathematical model is based on an extensive database, formed as a result of a comprehensive clinical and physiological examination and determination of the elemental composition of the biosubstrates of 910 persons of different sex and age. The determination of the content of chemical elements in biosubstrates was carried out by atomic-emission and mass-spectrometry methods. Using the Elan 9000 mass spectrometer (Perkin Elmer, USA), the content of As, Cd, Co, Cr, Cu, Fe, Mn, Ni, Pb, Se, Si, Sn, Ti, V was determined and the Optima 2000 V atomic emission spectrometer (Perkin Elmer, USA) was used to determine Ca, Mg, P, Zn, K, Na. In total, data on the content of 24 chemical elements were obtained. The evaluation of physical development was carried out according to recommendations adopted by WHO and uniform anthropometric methods [7] using standard measuring instruments. Anthropometric data were estimated from centile tables developed for this age category [8] and were used to calculate the body mass index (BMI). The results were interpreted according to the developed standards.

For descriptive reasons, the results are shown in figure 1 in relative units. In this case, the norm is reflected by a single line for all elements, the maximum and minimum possible change by two other lines located at a distance from the norm proportional to the parameter α from formula (5) (in this case, we choose lines with the values 0.5 and 1.5 as these lines).

5. Discussion

The results obtained show that patients 2 and 4 have an excess in content of Ca and Mg. Patient 2 exceeded the concentration of Zn, and patient 4 – Ti. In addition, Patient 5 has an excess of Se.

The lowered concentration of elements is much more widely presented. All patients have a lower concentration of Cd, K, Li, Na, Pb. In four patients out of five, Hg, I, Sn is not sufficient. Three of the five patients have a lower concentration of Cr and Mn, two of them – As, Co, Fe. Finally, patient 1 lacks copper and silicon, patient 5 lacks nickel. This way of presenting information makes it easy to analyze the dynamics of the correction of elemental status by comparing the two graphs before and after taking the drugs. For receiving more complete idea of the general regularities of accumulation or loss of chemical elements in an organism additional axes can be introduced in the multidimensional space, over which the values of physiological, anthropometric, metabolic and other indices are placed. Then assessing the dispersion of the concentrations of elements, one can find additional symptoms important for the doctor-diagnostician. For example, a small dispersion of the physiological index for some elements in a certain group of patients may indicate a strong correlation.

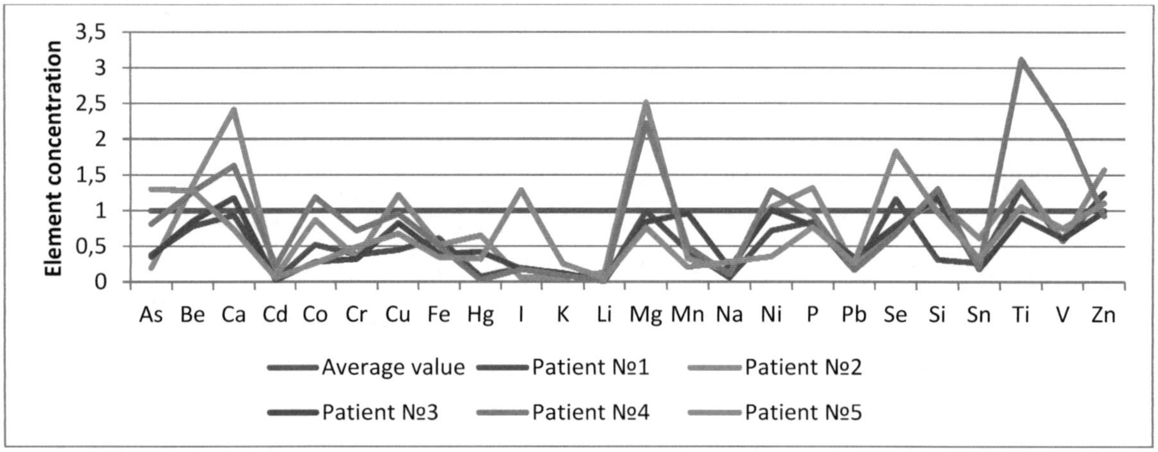

Figure 1. Graph of the relative concentration of elements for five patients

6. Conclusions

Thus, the multidimensional representation of information and the developed processing algorithm allow to increase the accuracy of the assessing of the elemental status of the organism. The results of this study allow us to create a system to support the assessing the elemental status of the organism, taking into account the ecologo-physiological adaptation. Corresponding algorithms and their software implementation make it possible to minimize the influence of the subjectivity on the accuracy of diagnosis and the formation of the most effective treatment strategy. Thus, the doctor's work is transferred to another level with a quick orientation in a large amount of information about the patient and a justified choice of the strategy for correcting the elemental status.

Acknowledgments

The current investigation was supported by the Government of the Orenburg region (project N32, 31.07.2017).

References

[1] Burtseva T I, Notova S V, Skalny A V and Zhivaev N G 2015 Modeling of the system of ecological monitoring for the state of selenium status of the population *Innovations and investments* **6** p 149–152

[2] Notova S V, Laryushina I E, Kiyaeva E V, Tsypin A P and Ermakova N V 2016 Elemental status of students of different social groups *Human Ecology* **12** p 43–48

[3] Notova S V, Alidzhanova I E, Kiyaeva E V and Akimov S S 2015 Indicators of psychophysiological adaptation of students of different social groups *Human Ecology* **11** p 41–47

[4] Yushkevich A P 1972 *The history of mathematics (volume 3). Mathematics of the XVIII century* (Moscow: Nauka)

[5] Pishukhin A M and Ahmadyanova G F 2015 Designing an educational route in the competence space *Vestnik of the Orenburg State University* **3** p 21–24

[6] Akhmedyanova G F et al 2013 Agent-oriented approach to modeling the learning process *Fundamental research* **11-3** p 521–524

IOP Conf. Series: Journal of Physics: Conf. Series **913** (2017) 012002 doi:10.1088/1742-6596/913/1/012002

[7] Martirosov E G and Rudnev S G 2002 Composition of the human body. New technologies and methods *Sport, medicine and health* **3** p 5–9
[8] Kuznetsova M V and Setko N P 2005 *Tables for assessing the physical development of boys and girls aged 17–21 in Orenburg: a manual for physicians* (Orenburg)

Big Data Analytics and Pattern Recognition Methods in the Problem of Optimization of Technological Processes in Metallurgical Production

D Gainanov[1] and D Berenov[2]

[1] The head of department "Big Data Analytics and Methods of Video Analysis", Ural Federal University, Ekaterinburg, Russia
[2] PhD student, Ural Federal University, Ekaterinburg, Russia

E-mail: damir.gainanov@gmail.com

Abstract. The paper considers the problem of predicting the quality of metallurgical production in a general formulation, when a huge amount of historical data of technological processes and product quality can be partitioned into a set of discrete classes of efficiency. Assuming that in the production process the technological process route can be changed, the problem of choosing the optimal route becomes important. When constructing the decision tree that recognizes the class of efficiency of the technological process, a special criterion for optimality of the partitioning of the set of classes of efficiency into two classes is introduced corresponding to the left and right branches of the decision tree in the node under consideration. The partitioning obtained by the proposed approach is close to optimal and can form the basis for constructing a decision tree, with the help of which a route is chosen to continue processing.

1. Introduction

The methods of Big Data analytics are widely used in various fields including the field of production optimization. Metallurgical production, due to a number of technical features, requires special attention to the quality of products. The problem of increasing the efficiency of production in terms of product quality can be investigated using a mathematical methods, since many routing processes can be described in terms of graph theory. At the same time, an adequate solution to the problem of predicting the quality of metallurgical production should take into account the historical experience of the production. Thus, the research of the problem covers not only the field of development and formalization of the mathematical model of the production process, but also the field of development of methods for analyzing and processing of Big Data.

In work [1] the problem of predicting the defects in metallurgical production was formalized and effective decision algorithms were obtained. This problem is a special case of the overall quality prediction problem and is investigated in this work using Big Data technologies. To solve this problem, methods such as an alternative covers or committees constructions, described, for example, in [2–6], can be used in the formulation of the research of infeasible systems. In this paper, we research the problem of predicting the quality of metallurgical production in a general formulation, which reduces to the problem of pattern recognition in a geometric formulation. The results in the solution of the problem of pattern recognition in a geometric formulation are given,

IOP Conf. Series: Journal of Physics: Conf. Series **913** (2017) 012003 doi:10.1088/1742-6596/913/1/012003

for example, in [7, 8]. In the present paper the formal justification and the mathematical formulation are given. Data on completed technological processes form a training sample of the supervised learning. In view of the large dimensionality of the training sample, the need for developing effective methods for analyzing Big Data follows in an obvious way. A special criterion is introduced for the optimal partition of the set of classes of efficiency of technological processes, which makes it possible to significantly reduce the computational costs of constructing a decision tree. A formal description of the algorithm and the results of the implementation on the test case are given.

2. Basic definitions

In the process of metallurgical production, a certain unit of production (PU) is processed sequentially on a number of technological aggregates. The methods of graph theory provide an adequate tool for modelling production processes, including metallurgical ones.

Definition 2.1 A directed graph $\vec{G}=(\mathrm{A},E)$, the set of vertices of which corresponds to a set of technological aggregates, is called an infrastructure graph, if $(A_1,A_2)\in E$ if and only if the output PU of the aggregate A_1 can serve as the input PU for the aggregate A_2.

The sequence of the aggregates forms a technological route (TR) for processing the PU. At the same time, according to the results of the processing, for each unit of the PU, values are assigned for a given set of parameters. In these terms, TR can be defined as a path in an infrastructural graph, that is, $P=\left(A_{i_1},A_{i_2},\ldots,A_{i_k}\right)$. We denote the set of all TRs by $\mathrm{P}=\{P_1,\ldots,P_k\}$.

Definition 2.2 Sequence

$$AI_i=\left(A_{i1},\mathrm{P}_{i1}\left(AI_i\right),\ldots,A_{is},\mathrm{P}_{is}\left(AI_i\right)\right),$$

where $\mathrm{P}_{ij}\left(AI_i\right)$ is a set of parameter values for a PU in a specific implementation of TR AI_i, is called a completed technological route (CTR).

A set of CTR is formed continuously in the process of production activity. Thus, the problem arises of processing historical data. In this case, in view of the fact that each PU is characterized by a set of parameters of large dimension, the problem arises of processing of Big Data. In this paper we propose an approach to solving the problem of optimization of technological processes using methods of graph theory and pattern recognition.

We denote by $G^k(A)$ the set of all vertices A of the graph $\vec{G}$ such that there exists a simple path from the vertex A to the vertex A' of the length $k-1$. The geometric meaning of this notation is close to the notion of a neighborhood of the k-th order in the graph.

Definition 2.3 Technological pyramid $\mathrm{Pir}(\vec{G},A)$ with a root vertex A is a subgraph of a graph $\vec{G}$, generated by a set of vertices $\langle A\cup G(A)\cup G^2(A)\cup\ldots\cup G^k(A)\rangle_{\vec{G}}$ such that any directed path P of the graph $\vec{G}$, starting at the vertex A, lies entirely in this subgraph.

Definition 2.4 A fork-vertex of a graph (subgraph) is a vertex $A\in\mathrm{A}$, such that $|G(A)|>1$.

Definition 2.5 A terminal vertex of a graph (subgraph) is a vertex, from which no arc leaves in this graph (subgraph).

Let A' be the set of terminal vertices of the technological pyramid $\mathrm{Pir}(\vec{G},A)$. For each terminal vertex $A_i\in\mathrm{A}'$ there is some PU, which is the output PU for this vertex, and there may be several EPs depending on the type of CTR, as a result of which the given PU has been obtained.

Definition 2.6 An CTR is called a productive if the PU at the output of the terminal vertex $term(AI_i)$ of the CTR AI_i is one of the types of the final product deliverable to the market.

The final PU obtained as a result of passing the productive CTR is characterized by a quality indicator. For example, defective or suitable parts determine the quality of the final PU.

Definition 2.7 The effectiveness of a productive CTR AI_i is the value calculated as

$$Ef(AI_j) = \frac{\mathrm{Price}(term(AI_j)) - \mathrm{C}(term(AI_j))}{\mathrm{C}(term(AI_j))}$$

where $\mathrm{Price}(term(AI_j))$ characterizes the market price of the unit of measurement of the final PU and $\mathrm{C}(term(AI_j))$ is cost of production of the unit of measurement of the final PU.

The investigated process of metallurgical production allows the choice of TR for the continuation of processing of PU. For different productive CTRs, the sequences of technological aggregates may coincide, however, the PU parameters are different for different productive CTRs, which entails a difference in efficiency indicators. Thus, the practical meaning of the problem is to choose such a TR to continue processing the PU, so that the final efficiency indicator (that is, the quality index of the final PU) will reach its optimal value. The basis of this choice, obviously, can be the historical experience accumulated in the production process.

3. The problem of forecasting the quality of production

Consider the fragment of the infrastructure graph $\vec{G} = (\mathrm{A}, E)$ presented in Figure 1.

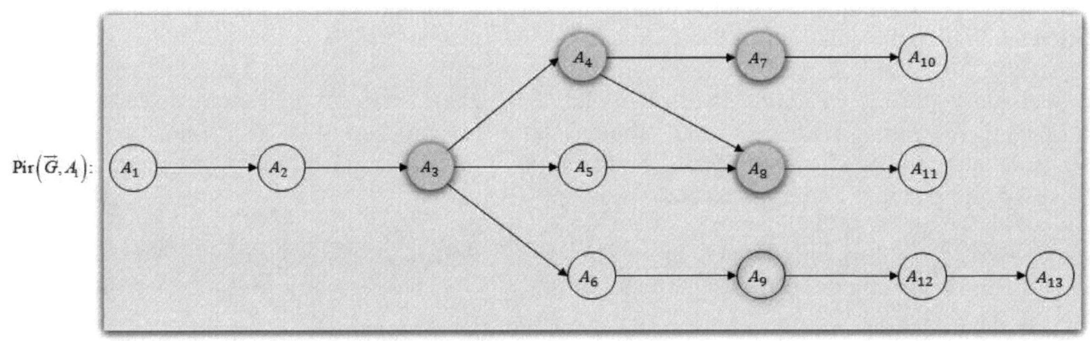

Figure 1. Technological pyramid $\mathrm{Pir}(\vec{G}, A_1)$ with a root vertex A_1

The fork-vertices in the represented technological pyramid are A_3, A_4. That is, with the reaching the aggregates A_3, A_4 TR can be changed in order to improve the quality of the expected unit of production. The set of TMs contained in the technological pyramid $\mathrm{Pir}(\vec{G}, A_1)$, corresponds to a set of directed paths:

$$P_1 = (A_1, A_2, A_3, A_4, A_7, A_{10}),$$

$$P_2 = (A_1, A_2, A_3, A_4, A_8, A_{11}),$$

$$P_3 = (A_1, A_2, A_3, A_5, A_8, A_{11}),$$

$$P_4 = (A_1, A_2, A_3, A_6, A_9, A_{12}, A_{13}).$$

Each CTR in the pyramid $\mathrm{Pir}(\vec{G}, A_1)$ follows a certain TR, and before reaching the fork-vertex some CTRs should follow the same TR. In other words, the CTR is a sequence of aggregates given together with the parameters for a unit of output for each vertex in the directed path. It is clear that for each CTR the parameters of the unit of production will differ even for the same aggregate. Thus, at each time-moment, a number of CTRs will be generated for each TR in the infrastructure graph. For the technological pyramid $\mathrm{Pir}(\vec{G}, A_1)$ under consideration, we will have the following types of productive CTRs:

$$AI_1 = (A_1, \mathrm{P}_{11}, A_2, \mathrm{P}_{12}, A_3, \mathrm{P}_{13}, A_4, \mathrm{P}_{14}, A_7, \mathrm{P}_{17}, A_{10}, \mathrm{P}_{110}),$$

$$AI_2 = (A_1, \mathrm{P}_{21}, A_2, \mathrm{P}_{22}, A_3, \mathrm{P}_{23} A_4, \mathrm{P}_{24}, A_8, \mathrm{P}_{28}, A_{12}, \mathrm{P}_{211}),$$

etc.

A set of productive CTRs are formed continuously in the stream mode at each control time-moment. At the same time, the number of parameters for each productive CTR can reach a dimension of several tens or even hundreds, and the number of CTR in the training sample can achieve several hundred and tens of thousands and more. Thus, the solution of the problem of predicting the expected quality of the product (the choosing of the TR for continuation of production) requires researches in the field of processing and analyzing of Big Data.

Let at some time-moment t a set of productive CTRs $\mathrm{P}_{\mathrm{CTR}}(A_0, t) = \{AI_i : i = [1, q]\}$ with an initial vertex A_0 be formed such that their sections up to a certain fork-vertex A' coincide in the part of the passage of the aggregates. We divide the set of all CTRs $\{AI_i : i = [1, q]\}$ into several classes

$$\mathrm{P}_{\mathrm{CTR}}(A_0, t) = \mathrm{P}^{(1)}(A_0, t) \dot\cup \mathrm{P}^{(2)}(A_0, t) \dot\cup \mathrm{P}^{(3)}(A_0, t) \dot\cup \ldots \dot\cup \mathrm{P}^{(k)}(A_0, t) \tag{1}$$

such that AI_i and AI_j belong to the same class if and only if $P_i = P_j$, that is, the productive CTRs belong to the same class, if they follow the same TR (they coincide in the part of the passage of the aggregates during the whole period of the route). In the example under consideration, we have a partition:

$$\mathrm{P}_{\mathrm{CTR}}(A_1, t) = \mathrm{P}^{(1)}(A_1, t) = \{AI_1^i\} \dot\cup \mathrm{P}^{(2)}(A_1, t) = \{AI_2^i\} \dot\cup$$

$$\dot\cup \mathrm{P}^{(3)}(A_1, t) = \{AI_3^i\} \dot\cup \mathrm{P}^{(4)}(A_1, t) = \{AI_4^i\},$$

while the productive CTRs AI^i and AI^j belonging to the same class differ only in the part of the parameter sets for each aggregate in the sequence. It is clear that for each class in the partition (1) there is a TR common to all productive CTRs in the class. Due to the fact that the TR which belongs to the pyramid $\mathrm{Pir}(\vec{G}, A_1)$, has a common section (A_1, A_2, A_3) up to the fork-vertex A_3, an applied problem of choosing a technological route arises in the process of its implementation. The substantive meaning of the problem is that for a specific TR implementation according to the initial problem it may turn out that further implementation of this TR may prove to be economically impractical, since the expected quality of the received PU will not be high enough at the planned costs for the implementation of this TR. Thus, the problem of predicting the quality of products

IOP Conf. Series: Journal of Physics: Conf. Series **913** (2017) 012003 doi:10.1088/1742-6596/913/1/012003

of metallurgical production is to determine which of the technological routes $P_i : i = [1, k]$, should be chosen for further follow-up when the fork-vertex A' is reached. To develop effective algorithms for choosing the optimal route for the continuation of production, we reduce the study to the problem of pattern recognition in a geometric formulation.

Let us consider the formulation of the problem. Consider the partition (1) of the CTR in the technological pyramid $\mathrm{Pir}\left(\vec{G}, A_0\right)$. For each class of productive CTRs $\mathrm{P}^{(i)}\left(A_0, t\right), i = [1, k]$, in the partition (1) we compile a sample

$$Z(P_i) : \left(Ef\left(AI_j\right) = \frac{\mathrm{Price}\left(term\left(AI_j\right)\right) - \mathrm{C}\left(term\left(AI_j\right)\right)}{\mathrm{C}\left(term\left(AI_j\right)\right)}, i, BI_j \right), \tag{2}$$

where BI_j is the CTR on the section up to the fork-vertex A' and $Ef\left(AI_j\right)$ is defined in Definition 2.7. If the set of values of efficiency $Ef\left(AI_j\right)$ is divided into several discrete intervals $E_1, E_2, \ldots, E_m$ then the prediction problem consists in determining into which class $E_1, \ldots, E_m$ the PU falls when the TR is continued in the class $i, i = [1, \mathrm{m}]$.

Each productive CTR in the sample (2) can be represented as a multidimensional vector

$$\mathbf{a}_i = \left(a_{i_0}, a_{i_1}, a_{j_1}, \ldots, a_{j_k}\right),$$

where $a_{i_0} \in \{E_1, \ldots, E_m\}$ is the value of the efficiency of the CTR, a_{i_1} is the TR identifier of this CTR, and the values of the variables $a_{j_1}, \ldots, a_{j_k}$ correspond to the aggregates and parameters of the PU in the CTR BI_i. Thus, for a given set of n - dimensional vectors, divided into m classes,

$$\mathrm{A} = \left\{\left(a_{i_1}, a_{i_2}, \ldots, a_{i_n}\right)\right\} = \mathrm{A}_1 \cup \mathrm{A}_2 \cup \ldots \cup \mathrm{A}_m,$$

it is required to construct a decision rule for assigning an arbitrary input vector $\mathbf{b}_i$ to one of the classes $\mathrm{A}_1, \mathrm{A}_2, \ldots, \mathrm{A}_m$.

Suppose that for each class of productive CTRs $\mathrm{P}^{(i)}\left(A', t\right), i = [1, k]$, a representative sample of historical CTRs has been compiled, for which the considered problem of forecasting the quality of products is solved while continuing the technological process in the i-th variant. Then the method of practical use of the proposed method is as follows. The current TR is executed before reaching the fork-vertex A'. As a result, we get CTR BI_s on the section $\left(A_0, A_1, \ldots, A'\right)$. Then, by consistently applying the found decision rules for each of the k classes for the CTR BI_s, we obtain sequences of values of the predicted value Ef for all variants of the continuation of the technological process after passing the fork-vertex A'. The obtained values Ef can be effectively applied for choosing the variant of the continuation of the technological process. For example, we can choose the variant of continuing the technological process for some Ef which takes the maximum value. More complex variants of the continuation of the technological process, can be constructed if we take into account the fulfillment of the planned indicators for the volumes of production of various types of products.

4. The algorithm for optimal partitioning of the set of classes for constructing the decision tree

In [1], a special case of the problem of predicting the quality of production is investigated. The set of values of the efficiency of productive CTRs (quality of the final PU) is divided into 2 classes of fit and defective parts. This formulation was named in the paper as the problem of predicting the defectiveness of the production where it is required to determine whether the final PU is fit if the current TR continues. The proposed approach is based on reducing the problem of pattern recognition in a geometric formulation to the research of a system of linear inequalities.

The more general formulation of the problem is considered in [9]. In this paper, an algorithm for constructing a decision function for the node of the decision tree $G=(V,E)$ is presented. The main idea of the algorithm is to achieve the maximum value of the separating function. The training set of vectors would be separated by a hyperplane

$$f(\mathbf{a})=\mathbf{a}\cdot n_v-\varepsilon_v=0,$$

where v is the vertex of the decision tree with the given sample A_v, and the vectors n_v,ε_v are given in such a way that the descendants v_1,v_2 of the vertex v in the decision tree correspond to the samples

$$A_{v_1}=\{\mathbf{a}_j\in A_v:\langle n_v,\mathbf{a}_j\rangle\leq\varepsilon_v\}\neq\varnothing,$$

$$A_{v_2}=\{\mathbf{a}_j\in A_v:\langle n_v,\mathbf{a}_j\rangle>\varepsilon_v\}\neq\varnothing.$$

Reaching maximum of the value

$$\operatorname{discrim}(A_v,n_v,\varepsilon_v)=\sum_{i\in I}\left|\frac{|A_{v_1}\cap A_i|}{|A_{v_1}|}-\frac{|A_{v_2}\cap A_i|}{|A_{v_2}|}\right|$$

provides an effective partition of the set of vectors according to the classes of the training sample. Thus, the practical implementation of the approach requires the choice of partition A_v and vectors n_v,ε_v such that the value $\operatorname{discrim}(A_v,n_v,\varepsilon_v)$ reaches its maximum.

In this paper, we present an algorithm for partitioning a sample A_v into two classes $I=I_1\dot{\cup}I_2$, an important advantage of which is a significant reduction in the amount of computation when choosing the optimal partition. The result of the implementation of the developed algorithm serves as a basis for constructing a decision tree, while the choice of vectors n_v,ε_v for assigning an arbitrary vector to a certain class of training sample can be carried out using the algorithm proposed in [9].

Let us assume that a sample $\mathrm{A}=\mathrm{A}_1\dot{\cup}\mathrm{A}_2\dot{\cup}\ldots\dot{\cup}\mathrm{A}_m$ is given, a set of classes $I=[1,m]$ and $I_i=\{j:\mathbf{a}_j\in\mathrm{A}_i\}$. We introduce the notation:

$$a_{ij}=\sum\{(\mathbf{a}_{k_1}-\mathbf{a}_{k_2}):\mathbf{a}_{k_1}\in\mathrm{A}_i,\mathbf{a}_{k_2}\in\mathrm{A}_j\},i,j\in[1,k].$$

$$a(I_1,I_2)=\sum_{i\in I_1,j\in I_2}a_{ij}. \tag{3}$$

Among all partitions of the form

$$I_1,I_2\subseteq[1,m]:I_1,I_2\neq\varnothing,I_1\dot{\cup}I_2=I,$$

for the sample A_v given for the vertex v of the decision tree $G=(V,E)$, it is required to choose such one that the value $\|a(I_1,I_2)\|$ reaches its maximum, that is

$$\|a(I_1,I_2)\| \to \max . \tag{4}$$

Suppose that for a training sample $A = A_1 \dot{\cup} \ldots \dot{\cup} A_m$ the sum of differences of the form (3) is calculated pairwise for all classes. To search for a partition $I = I_1 \dot{\cup} I_2$, which provides a maximum for the function $a(I_1,I_2)$, we would consider the following heuristic algorithm.

Input data: the set of classes $I=[1,\mathrm{m}]$

Output data: the partition $I = I' \dot{\cup} I"$

Let $I' = \varnothing, I" = I$

While $I" \neq \varnothing$ **do**

$i_0 = \arg\max_{i\in I"}\left(\left\{\| a\left(I' \dot{\cup}\{i\}, I",\ \{i\}\right) \|\right\}\right)$

$I' \leftarrow I' \dot{\cup} \{i_0\}$

$I" \leftarrow I",\ \{i_0\}$

End of condition

End of algorithm

In the process of the algorithm, it is constructed a sequence of partitions of the form

$$I_1 = \varnothing \dot{\cup} I ,$$

$$I_2 = \{i_1\} \dot{\cup} I ,\ \{i_1\},$$

$$\ldots\ldots\ldots\ldots ,$$

$$I_n = I \dot{\cup} \varnothing ,$$

among which one should choose the such one that $\|a(I', I")\|$ is the maximal. Thus, for a given training sample and a set of classes, a partition is constructed taking into account the maximization of the value $\|a(I_1,I_2)\|$. In this case, the partition is the basis for constructing a decision tree for classifying vectors from the training sample.

Let's consider an example where the search for a solution is carried out using the developed approach.

Example 4.1 Let there be given a set $I=\{1,2,3,4\}$ and 4 vectors a_1,a_2,a_3,a_4.

Then the matrix of values for a_{ij} will have the form

$$a_{ij} = \left\| \begin{array}{c|cccc} & 1 & 2 & 3 & 4 \\ 1 & 0 & (1,0) & (0,-1) & (1,-1) \\ 2 & (0,-1) & 0 & (-1,-1) & (0,-1) \\ 3 & (0,1) & (1,1) & 0 & (1,0) \\ 4 & (-1,1) & (0,1) & (-1,0) & 0 \end{array} \right\| .$$

IOP Conf. Series: Journal of Physics: Conf. Series **913** (2017) 012003 doi:10.1088/1742-6596/913/1/012003

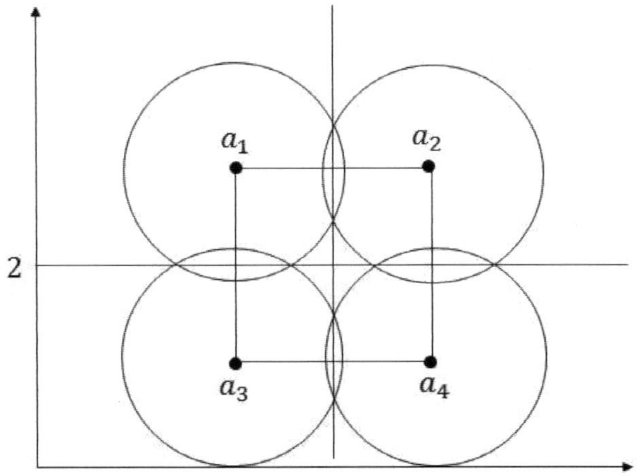

Figure 2. A partitioning the set of vectors into classes

Using the developed approach, we construct the optimal partition of vectors into classes.

1. $I_1 = \{1\}, I_2 = \{2,3,4\}$,

   $a(I_1, I_2) = a_{12} + a_{13} + a_{14} = (1,0) + (0,-1) + (1,-1) = (2,-2)$,

   $\| a(I_1, I_2) \| = \sqrt{4+4} = 2\sqrt{2} \quad 2{,}8$.

2. $I_1 = \{1,2\}, I_2 = \{3,4\}$,

   $a(I_1, I_2) = a_{13} + a_{14} + a_{23} + a_{24} = (0,-1) + (1,-1) + (-1,-1) + (0,-1) = (0,-4)$,

   $\| a(I_1, I_2) \| = 4$.

3. $I_1 = \{1,2,3\}, I_2 = \{4\}$,

   $a(I_1, I_2) = a_{14} + a_{24} + a_{34} = (1,-1) + (0,-1) + (1,0) = (2,-2)$,

   $\| a(I_1, I_2) \| = \sqrt{4+4} = 2\sqrt{2} \quad 2{,}8$.

The maximal value $a(I_1, I_2)$ is achieved for the partition 2, that is

$$I = \{1,2\} \dot{\cup} \{3,4\},$$

which looks like a logical result in a practical sense.

The proposed search algorithm for optimal partitioning is effective due to the use of a matrix $| a_{ij} |_1^k$ with previously calculated values a_{ij}. When processing huge amount of historical data, where the sample size can achieve hundreds of thousands and millions of vectors, it is advisable to apply the method proposed in the present work not to the initial sample, but to a sample obtained from the initial one as a result of sample clustering, for example, using the effective VNS-clustering method from [10] with the replacement of each cluster by the n - dimensional vector representing it.

5. Conclusion

The paper considers the applied problem of optimization of technological processes in metallurgical production. Historical data generated by the implementation of productive CTRs forms the technological database, and, within the framework of the developed approach, a training sample of the supervised learning. An approach is proposed to solve the problem of clustering the set of training sample vectors, which allows to significantly narrow the solution search space and reduce computational costs. The main idea of the approach is the introduction of a special criterion for the optimality of the partition, on the basis of which an algorithm for search a solution close to the optimal has been developed. The algorithm is implemented in a test case, and the results obtained are the basis for constructing a decision tree for the initial classification problem.

References

[1] Gainanov D N and Berenov D A 2017 Algorithm for predicting the Quality of the product of Metallurgical Production *Proc. Int. Conf. on Big Data and Advanced Analytics* (Minsk) p 65-70

[2] Gainanov D N 2014 *Combinatorial Geometry and Graphs in the Analysis of Infeasible Systems and Pattern Recognition* (Moscow: Nauka)

[3] Gainanov Damir N 2016 Graphs for Pattern Recognition. *Infeasible Systems of Linear Inequalities* (DeGruyter)

[4] Gainanov D N 1985 On combinatorial properties of infeasible systems of linear inequalities and convex polyhedra *Math notices* vol 38 3 p 463–474

[5] Mazurov Vl D 1990 *Committees method in problems of optimization and classification* (Moscow: Nauka)

[6] Mazurov Vl D and Khachai M Yu 2004 Committees of systems of linear inequalities *Automation and Remote Control* 2 p 43–54

[7] Gainanov D N 1992 *Alternative Covers and Independence Systems in Pattern Recognition Pattern Recognition and Image Analysis* vol 2 **2** p 147–160

[8] Gainanov D N and Matveev A O 1991 Lattice Diagonals and Geometric Pattern Recognition Problems *Pattern Recognition and Image Analysis* vol 1 **3** p 277–282

[9] Gainanov D N and Berenov D A 2017 *Algorithm for Predicting the Quality of the Product of Metallurgical Production Proc. Int. Conf. on Optimization and Applications* (Pertovac) accepted for publication

[10] Hansen P and Mladenovic N 2001 J-MEANS: a new local search heuristic for minimum sum of squares clustering *Pattern Recognition* **34** p 405–413

IOP Conf. Series: Journal of Physics: Conf. Series **913** (2017) 012004 doi:10.1088/1742-6596/913/1/012004

Deep Learning for ECG Classification

B Pyakillya, N Kazachenko and N Mikhailovsky

Tomsk Polytechnic University, NTR lab

e-mail: bpakilla@sinergo.ru, nkazachenko@sinergo.ru, nickm@ntrlab.com

Abstract. The importance of ECG classification is very high now due to many current medical applications where this problem can be stated. Currently, there are many machine learning (ML) solutions which can be used for analyzing and classifying ECG data. However, the main disadvantages of these ML results is use of heuristic hand-crafted or engineered features with shallow feature learning architectures. The problem relies in the possibility not to find most appropriate features which will give high classification accuracy in this ECG problem. One of the proposing solution is to use deep learning architectures where first layers of convolutional neurons behave as feature extractors and in the end some fully-connected (FCN) layers are used for making final decision about ECG classes. In this work the deep learning architecture with 1D convolutional layers and FCN layers for ECG classification is presented and some classification results are showed.

1. Introduction

Most part of heart problems, like Myocardial Infarction, AV Block, Ventricular Tachycardia and Atrial Fibrillation etc. can all be diagnosed from ECG signals with an estimated hundreds of millions ECGs recorded annually [1, 2]. We investigate the task of arrhythmia detection from a single short ECG lead. This is known to be a challenging task for computers but can usually be determined by an expert from a single, well-placed lead.

The importance of ECG classification is very high now due to many current medical applications where this problem can be stated. Currently, there are many machine learning (ML) solutions which can be used for analyzing and classifying ECG data. However, the main disadvantages of these ML results are use of heuristic hand-crafted or engineered features with shallow feature learning architectures. The problem relies in the possibility not to find most appropriate features which will give high classification accuracy in this ECG problem. One of the proposing solution is to use deep learning (DL) architectures where first layers of convolutional neurons behave as feature extractors and in the end some fully-connected (FCN) layers are used for making final decision about ECG classes. In this work the deep learning architecture with 1D convolutional layers and FCN layers for ECG classification is presented and some classification results are showed.

2. Current results in ECG Classification

As stated before, ECG classification problem is very important and had a strong attention from not only medical community but also community of computer scientists especially from AI field [1, 2]. The latest results were obtained from Andrew Ng's scientific group (Pranav Rajpurkar, Awni Y. Hannun, Masoumeh Haghpanahi, Codie Bourn), where they used 34-layer convolutional neural network and exceed the average cardiologist performance in both recall and precision [2].

One of the crucial property of the latter work was use of a dataset with more than 500 times the number of unique patients than other well-known corpora. Other works present two-class ECG classification algorithms with simple neural network architecture (ischemic heart disease or normal sinus rhythm), which works well but in case of low amount of data and special problem statement [3–5]. Unfortunately, mostly researchers don't have a possibility to use such kind of dataset due to many reasons. Obvious solution for this is to use of data augmentation technics, which are well fitted to time series classification problems [6–9]. Further, we will describe the available medical data and propose the classification algorithm.

3. Problem statement and data

This current ECG classification task can be described as task to decide to which class patient's ECG can be assigned. The number of classes are four: normal sinus rhythm, arrhythmic, other kind of rhythm, very noisy [10]. The given data is non-balanced, most part of belongs to normal sinus rhythm (60% of data). The Figure 1 presents the brief summary of the available data.

There were used different operations for creating balanced dataset like multiplication of existing ECG for some class by means of shifting time values. In addition there was a try to create some unified length of ECG by means of duplication time-series values.

Type	# recording	Time length (s)				
		Mean	SD	Max	Median	Min
Normal	5154	31.9	10.0	61.0	30	9.0
AF	771	31.6	12.5	60	30	10.0
Other rhythm	2557	34.1	11.8	60.9	30	9.1
Noisy	46	27.1	9.0	60	30	10.2
Total	**8528**	**32.5**	**10.9**	**61.0**	**30**	**9.0**

Figure 1. The Data Summary

It is crucial to look into ECG data which can be obtained from patients and decide what kind of preprocessing and machine learning algorithm we have to use. Firstly, most ECG data is time series data with duration about 30 or 60 s with sampling time about 0.003 s. There are many machine learning algorithms with some appropriate choice of features which deal with time series data. But as it stated in introduction the work presents use of feature extractors – convolutional neural networks (CNN) in special case (Figure 2). This feature extraction should free the developers from use of expert knowledge and hand-crafted features. It is not remove expert knowledge at all from development process but decrease the time to show the first prototype of ECG classification algorithm.

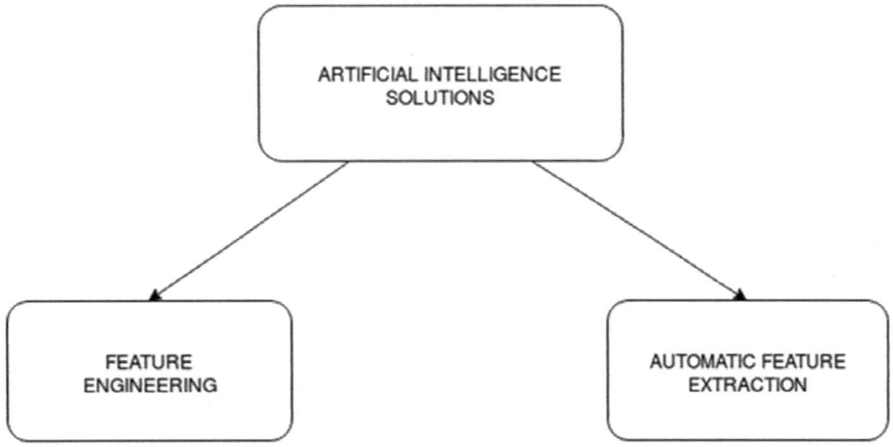

Figure 2. The simplified structure of AI solutions

This special case of ECG differs from usual one where CNN is used – image recognition task. In the latter case, data is always presented as 2D data with some color channels in contrary to time series where usually 1D data is used (Figure 3).

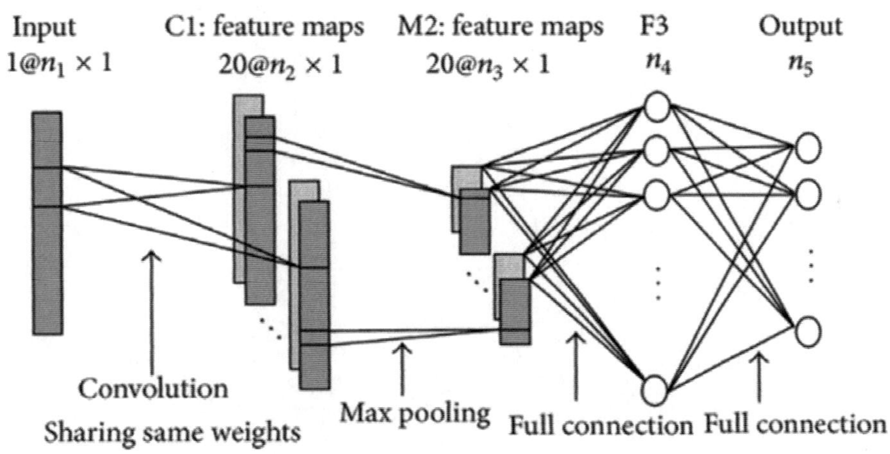

Figure 3. The example of CNN for Time series

In any case, CNN with appropriate architecture, which depends on data dimensionality and its structure, can give high accuracy in classification task. In current case, we use CNN 1D with the following architecture: 7 convolutional layers with filter width 5 and 128 neurons + max-pooling and dropout after every layer + GlovalAveragePooling + 3 FCN layers with (256/128/64) neurons + dropout after every layer + softmax layer with 4 outputs.

Also, there was used some standard preprocessing procedure like subtraction the mean from time-series values and division by standard deviation value, the batch size was equal to 256. All parts of this algorithm was implemented by means of Keras framework [11].

The Figure 4 and Figure 5 present the loss and accuracy values for the training steps consequently, the same works for Figure 6 and Figure 7 which show results for validation step.

IOP Conf. Series: Journal of Physics: Conf. Series **913** (2017) 012004 doi:10.1088/1742-6596/913/1/012004

In all computational experiments we use GPU NVIDIA GeForce GTX 1080 and CPU i5 – 6600 3.3 GHz. The obtained results can be compared to existed ones which were gained in other previous works in terms of precision.

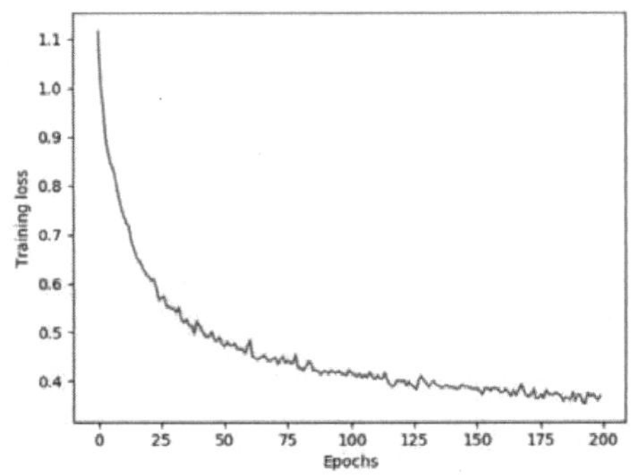

Figure 4. Plot of logarithmic loss for training step

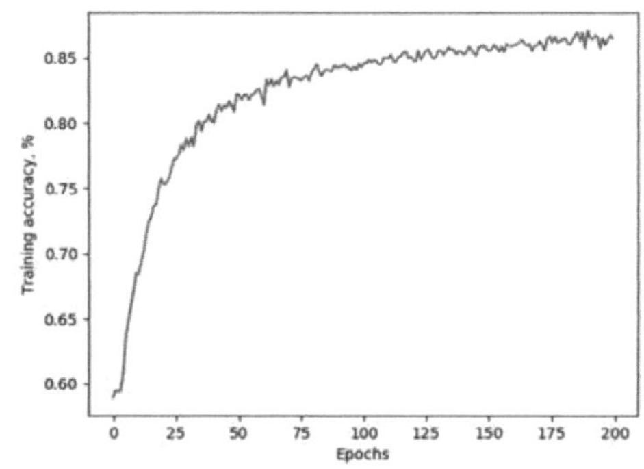

Figure 5. Plot of accuracy for training step

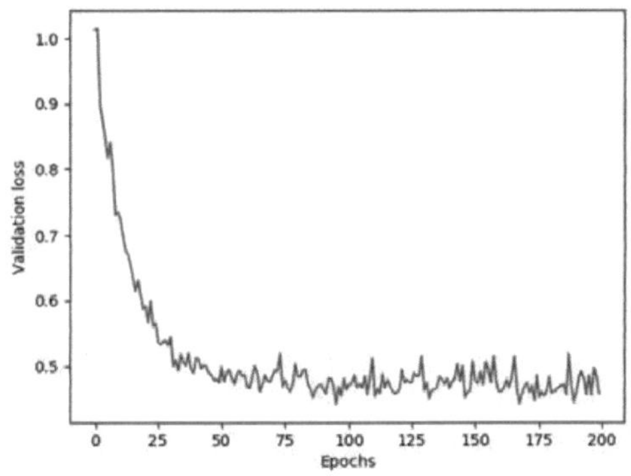

Figure 6. Plot of logarithmic loss for validation step

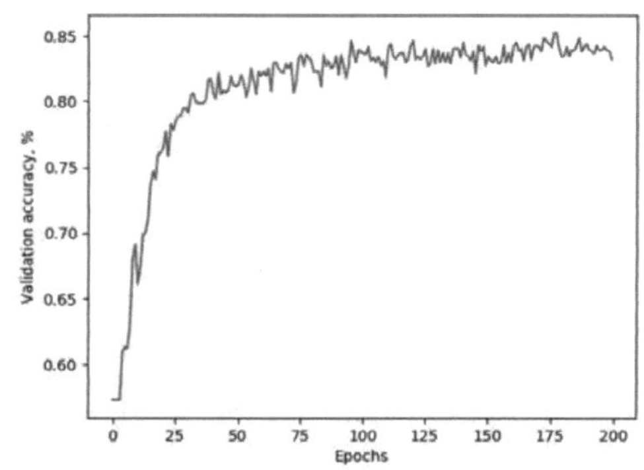

Figure 7. Plot of accuracy for validation step

4. Discussion

The obtained classification results show that the proposed DL solution can be used in task of ECG classification in the case when training data is unstructured, unbalanced and can be represented as 1D time series with standard time duration for portable single lead ECG devices. Also the algorithm's ability to extract features is very useful in the cases when there is no any medical or related subject specialists for feature engineering due to some reasons. Disadvantages of the work can be described as unsatisfied comparison to other DL solutions in terms of computational complexity, proposed architecture and optimization methods. Also, it is interesting to test a recurrent neural net (RNN, LSTM etc.) in ECG classification task because of its implicit ability to work with historical data like time series. In any case the main open problem is to decide exactly what kind of architecture should be used for given datasets – what number of neurons, layers and type of optimization method.

5. Conclusions

This work presented some ECG classification results about use of 1D convolutional neural networks with FCN layers on preprocessed time-series data. The best resulted accuracy on validation data is about 86%. In future there can be some upgrade in accuracy by means of some other preprocessing

IOP Conf. Series: Journal of Physics: Conf. Series **913** (2017) 012004 doi:10.1088/1742-6596/913/1/012004

procedures and creating more balanced dataset by use of GAN networks. In any case, the main goal is to present and use some machine learning algorithm without any feature engineered procedure and competing classification accuracy in comparison to human results.

References

[1] Heden B, Ohlsson, M et al 1996 Detection of frequently overlooked electrocardiographic lead reversals using artificial neural networks *The American journal of cardiology* **78 (5)** p 600–604

[2] Rajpurkar P, Hannun Awni Y, Haghpanahi M, Bourn C and Ng Y A 2017 Cardiologist-Level Arrhythmia Detection with Convolutional Neural Networks (*Preprint* arXiv:1707.01836v1)

[3] El-Khafif S H and El-Brawany M A 2013 Artificial Neural Network-Based Automated ECG Signal Classifier *ISRN Biomedical Engineering* **2013**(261917) 6 p

[4] Weems A, Harding M and Choi A 2016 Classification of the ECG Signal Using Artificial Neural Network *Proceedings of the 3rd International Conference on Intelligent Technologies and Engineering Systems* (ICITES2014) p 545–555

[5] Gupta A, Thomas B, Kumar P, Kumar S and Kumar Y 2014 Neural Network based indicative ECG classification *5th International Conference – Confluence The Next Generation Information Technology Summit* (Noida, 2014) p 277–279

[6] Jianbo Y, Nguyen M. N, San P. P, Li X and Krishnaswamy S 2015 Deep Convolutional Neural Networks on Multichannel Time Series for Human Activity Recognition *Proceedings of the Twenty-Fourth International Joint Conference on Artificial Intelligence* (IJCAI 2015) AAAI Press 3995-4001.

[7] Wong S C, Gatt A, Stamatescu V, McDonnell M D 2016 Understanding data augmentation for classification: when to warp? (*Preprint* arXiv:1609.08764v2)

[8] Le Guennec A, Malinowski S and Tavenard R 2016 Data Augmentation for Time Series Classification using Convolutional Neural Networks *ECML/PKDD Workshop on Advanced Analytics and Learning on Temporal Data* (Sep 2016, Riva Del Garda, Italy) 8 p

[9] Salamon J and Bello J P 2017 Deep Convolutional Neural Networks and Data Augmentation for Environmental Sound Classification *IEEE Signal Processing Letters* **24 (3)** p 279–283

[10] *AF Classification from a short single lead ECG recording: the PhysioNet/Computing in Cardiology Challenge 2017* Avaliable from: https://www.physionet.org/challenge/2017/

[11] *Keras: The Python Deep Learning library* Avaliable from: https://keras.io/

IOP Conf. Series: Journal of Physics: Conf. Series **913** (2017) 012005 doi:10.1088/1742-6596/913/1/012005

Use of machine learning methods to classify Universities based on the income structure

Alexandra Terlyga[1] and Igor Balk[2]

[1] Ural Federal University after the first President of Russia B.N. Eltsin, 61 Komsomolskaya st., Ekaterinburg, Russian Federation.
[2] Global Innovation Labs LLC, 258 harvard Str #352 Brookline MA 02446 USA

E-mail: info@innovationlabs.net

Abstract. In this paper we discuss use of machine learning methods such as self organizing maps, k-means and Ward's clustering to perform classification of universities based on their income. This classification will allow us to quantitate classification of universities as teaching, research, entrepreneur, etc. which is important tool for government, corporations and general public alike in setting expectation and selecting universities to achieve different goals.

1. Introduction

In the modern society Universities become the key factor of economic development, because they function not only as establishments training staff and leading scientific research but also become active participants of innovation processes in Economics [1]. University should make its contribution in development of regional and national economics by creating the system of transfer of the research results into industry, Etzkowitz named it "Triple Helix system of innovation".

From the end of 1990s much attention to the concept of entrepreneurial university is given by Academic society, which tries to define this phenomenon in its works [2–9]. In the abovementioned works the definition of entrepreneurial university is given with the help of qualitative characteristics set determining the university activity in the sphere of business studies, start-up formation and organization of the projects with enterprises on technology transfer to industry providing.

Thus problems of universities financial activity become important subject for consideration as they help to assess incomes structure with the purpose to find out the most valuable sources and taking management decisions in the direction of university to the entrepreneurial type. What criteria should be concerned while university entrepreneurial activity assessing? How during analyzing the structure of University incomes, development can be managed?

In the number of modern works there is an attempt to analyze university financial indicators and to promote offers on changes of the current state and structural alterations. Concepts and mechanisms of financial management by universities need innovations in connection with changes in the system of market economy and corresponding financial crisis [10]. Analysis of the situation with expenses at the University supposes forming the solution on budget management strengthening, in particular on financial assessment indicators system creation [11] Issues on approach to efficiency assessment of governmental means into research university infrastructure, regional innovation systems and public infrastructures are discussed [12]. Current condition of university finance management in Czech Republic is analyzed and offers on universities budget assignations allocation are performed in connection with required goals at the State University [13]. Comparing structures of incomes

IOP Conf. Series: Journal of Physics: Conf. Series **913** (2017) 012005 doi:10.1088/1742-6596/913/1/012005

and expenditures of Canadian and Ukrainian Universities offers on investments from regional and municipal budgets are represented [14]. The role of accounting information and its influence on organizational changes from the viewpoint of competences and staff responsibilities becomes more valuable [15]. Analysis of budgeting system in four different universities from four ethnical groups allows to make recommendations on accounting cultural diversity in education [16].

In this paper we will concentrate on study of university classification using machine learning clustering algorithms.

2. Methodology

In this study we used list of universities according to their ranking by U.S. News Best Global Universities. This ranking is mainly focused on the academic research done in the University and worldwide reputation. For the ranking it utilize the daft by Clarivate Analytics (earlier Thomson Reuters), and accordingly Web of Science indexing service [17, 18].

To obtain information about income sources we used data provided in consolidated reports of the US Universities. These reports have uniform structure appropriate for performing further structuring, compilation and clustering.

The original US News ranking consisted of 170 Universities. The first hundred were chosen for this study. To unify reports analysis of universities income items was performed and the structure, which most relevantly describes major articles of universities income was created.

Incomes from educational activity are divided into revenues from students training in major education programs and earnings from additional educational programs:

1) Net student Income
2) Continuing education and executive programs
3) Total sponsored support is divided into the following categories:
4) Federal grants
5) State, Local grants
6) Private grants
7) Recovery of indirect costs and allowances

Total research revenues is divided into

1) Research revenues
2) Special laboratories revenue

Revenues from enterprises founded to provide supplementary university infrastructure activity are presented in the item:

- Auxiliary enterprises

Gifts for current use and other types of charity are shown in the item:

- Contributions and gifts

The following items present investment activities of the Universities:

1) Endowment returns made available for operations
2) General Operation Activities(GOA) returns made available for operations
3) Other investment income

Other incomes which do not have uniform structure and depend on peculiarities of the university activity and which are shown in the report with the aggregative line, presented in the item:

- Other incomes.

According to the chosen structure of university revenues the unified database was established. Data was additionally normalized:

1) to the total number of students educated at the University.
2) to the total income of the University.

3. Overview of the clustering algorithms

In this study we have used Ward's Hierarchical Agglomerative Clustering Method, K-means Method and Self-organizing map (SOM).

IOP Conf. Series: Journal of Physics: Conf. Series **913** (2017) 012005 doi:10.1088/1742-6596/913/1/012005

First we have established existence of clusters using SOM. The Self-Organizing Map (SOM), commonly also known as Kohonen network [19, 20] is a computational method for the visualization and analysis of high-dimensional data, especially experimentally acquired information.

The Self-Organizing Map defines an ordered mapping, a kind of projection from a set of given data items onto a regular, usually two-dimensional grid. A model mi is associated with each grid node. These models are computed by the SOM algorithm. A data item will be mapped into the node whose model which is most similar to the data item, e.g., has the smallest distance from the data item in some metric.

Algorithm

1. Randomize the map's nodes' weight vectors
2. Grab an input vector D(t), which is a target input data vector.
3. Traverse each node in the map
    1. Use the Euclidean distance formula to find the similarity between the input vector and the map's node's weight vector
    2. Track the node that produces the smallest distance (this node is the best matching unit, BMU)
4. Update the nodes in the neighborhood of the BMU (including the BMU itself) by pulling them closer to the input vector

$$W_v(s+1) = W_v(s) + \theta(u,v,s) * \alpha(s) * (D(t) - W_v(s))$$

where

s – is the current iteration

λ – is the iteration limit

t – is the index of the target input data vector in the input data set D

D(t) – is a target input data vector

v – is the index of the node in the map

W_v – is the current weight vector of node v

u – is the index of the best matching unit (BMU) in the map

θ(u, v, s) – is a restraint due to distance from BMU, usually called the neighborhood function, and

α (s) – is a learning restraint due to iteration progress.

5. Increase s and repeat from step 2 while $s < \lambda$

Later we used Ward's clustering and k-means algorithm to confirm stability of the obtained clustering model and study it's structure.

Ward's Hierarchical Agglomerative Clustering Method was first introduced by Ward [21]. He suggested an agglomerative hierarchical clustering algorithm with the criterion for choosing the pair of clusters to merge at each level is defined by the optimal value of an objective function.

We define a distance as a positive, definite, symmetric mapping of a pair of observation vectors onto the positive real, which in addition satisfies the triangular inequality.

For observations i, j, k we have:

$$d(i,j) > 0; d(i,j) = 0 \Longleftrightarrow i = j;\ d(i,j) = d(j,i); d(i,j) \leq d(i,k) + d(k,j).$$

For an observation set, I, with i,j,k $\in$ I we can write the distance as a mapping from the Cartesian product of the observation set into the positive real : $d : I \times I \rightarrow R+$.

At each step of the clustering algorithm we find a pair of clusters that leads to a minimum increase in total within cluster variance after the merging i.e. weighted squared distance between cluster centers $(1/|q| \sum_{i \in q} d^2(i,q^*))$.

Initially all clusters contain only a singe point. We can define Euclidean distance squared using norm ||.||: if $i,i' \in R|J|$, i.e. these observations have values on attributes $j \in \{1, 2, \ldots, |J|\}$, J is the attribute set, |.| denotes cardinality, then

$$d^2(i,i') = \|i - i'\|2 = \sum_j (i_j - i'_j)^2$$

Suppose that clusters C_i and $C_j$were next to be merged. At this point all of the current pairwise cluster distances are known. The recursive formula gives the updated cluster distances following the pending merge of clusters C_i and C_j. Let

•d_{ij}, d_{ik} *and* d_{jk} be the pairwise distances between clusters C_i, C_j *and*C_k, respectively,

• $d_{(ij)k}$be the distance between the new cluster $C_i \cup C_j$*and* C_k.

An algorithm belongs to the Lance-Williams family if the updated cluster distance $d_{(ij)k}$ can be computed recursively by

$$d_{(ij)k} = \alpha_i d_{ik} + \alpha_j d_{jk} + \beta d_{ij} + \gamma |d_{ik} - d_{jk}|,$$

where $\alpha_i, \alpha_j, \beta$ *and* γ are parameters, which may depend on cluster sizes, that together with the cluster distance function d_{ij}determine the clustering algorithm.

The term "k-means" was first used by James MacQueen in 1967 [22] though the idea goes back to Hugo Steinhaus in 1957 [23]. The standard algorithm was first proposed by Stuart Lloyd in 1957 as a technique for pulse-code modulation, though it wasn't published outside of Bell Labs until 1982 [24].

Given a set of observations $(x_1, x_2, \ldots, x_n)$, where each observation is a d-dimensional real vector, k-means clustering aims to partition the n observations into k ($\leq$ n) sets S = $\{S_1, S_2, \ldots, S_k\}$ so as to minimize the within-cluster sum of squares (WCSS) (i.e. variance). Formally, the objective is to find:

$$\arg\min_S \sum_{i=1}^{k} \sum_{x \in S_i} \|x - \mu_i\|^2 = \arg\min_S \sum_{i=1}^{k} |S_i| \mathrm{Var} S_i$$

where μ_i is the mean of points in S_i. This is equivalent to minimizing the pairwise squared deviations of points in the same cluster:

$$\sum_{\text{Cluster } C_i} \sum_{\text{Dimension } d} \sum_{x,y \in C_i} (x_d - y_d)^2$$

Because the total variance is constant, this is also equivalent to maximizing the squared deviations between points in different clusters (between-cluster sum of squares, BCSS) [25].

Given an initial set of k means $m_1^{(1)}, \ldots, m_k^{(1)}$, the algorithm proceeds by alternating between two steps [26].

Assignment step: Assign each observation to the cluster whose mean yields the least within-cluster sum of squares (WCSS). Since the sum of squares is the squared Euclidean distance, this is intuitively the "nearest" mean.

$$S_i^{(t)} = \left\{x_p : \left\|x_p - m_i^{(t)}\right\|^2 \leq \left\|x_p - m_j^{(t)}\right\|^2 \ \forall\, j; 1 \leq j \leq k\right\}$$

where each x_p is assigned to exactly one $S^{(t)}$, even if it could be assigned to two or more of them.

Update step: Calculate the new means to be the centroids of the observations in the new clusters.

$$m_i^{(t+1)} = \frac{1}{\left|S_i^{(t)}\right|} \sum_{x_j \in S_i^{(t)}} x_j$$

Since the arithmetic mean is a least-squares estimator, this also minimizes the within-cluster sum of squares (WCSS) objective.

The algorithm has converged when the assignments no longer change. Since both steps optimize the WCSS objective, and there only exists a finite number of such partitioning's, the algorithm must converge to a (local) optimum. There is no guarantee that the global optimum is found using this algorithm.

The algorithm is often presented as assigning objects to the nearest cluster by distance. The standard algorithm aims at minimizing the WCSS objective, and thus assigns by "least sum of squares", which is exactly equivalent to assigning by the smallest Euclidean distance.

4. Empirical results

As a first step of our investigation we created 3 x 3 SOM so see the structure of the data and that cluster analysis will make sense in this case. As can be seen from the Figure 1 this approach worked very well and we have clear separation of two large groups of the universities (cells with centers around (0, 175), (1, 175) on one side and (1, 0), (2, 0) and (2.5, 1)). Now we can use k-means and Ward's clustering algorithm to validate our approach.

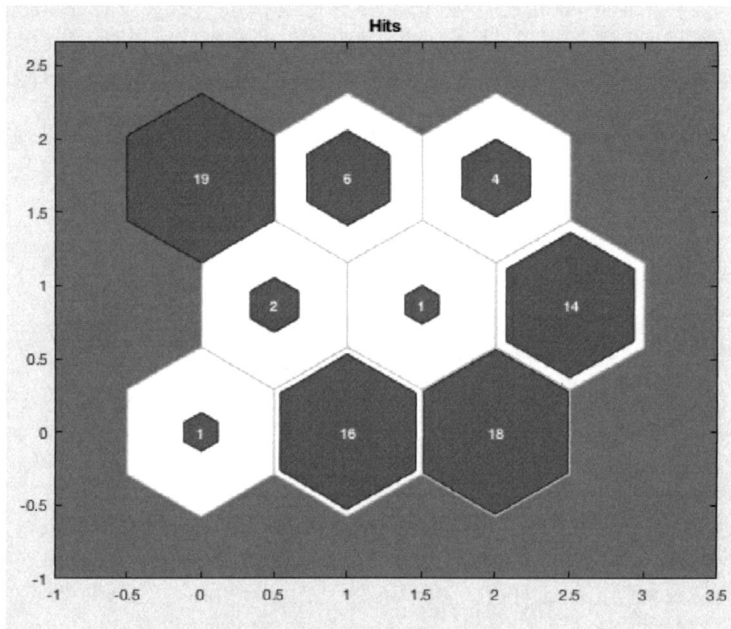

Figure 1. SOM hits with 3 x 3 mapping

As a check of validity of our clustering model we used silhouette metric. The silhouette value for each point is a measure of how similar that point is to points in its own cluster, when compared to points in other clusters. The silhouette value for the i-th point, Si, is defined as Si = (bi-ai)/ max(ai,bi) where ai is the average distance from the i-th point to the other points in the same cluster as i, and bi is the minimum average distance from the i-th point to points in a different cluster, minimized over clusters. The silhouette value ranges from -1 to +1. A high silhouette value indicates that i is well-matched to its own cluster, and poorly-matched to neighboring clusters. If most points have a high silhouette value, then the clustering solution is appropriate. If many points have a low or negative silhouette value, then the clustering solution may have either too many or too few clusters. The silhouette clustering evaluation criterion can be used with any distance metric. On the figures 2 and 3 we can see silhouette plot of k-means and Ward's clustering. We used simple squared Euclidean distances to calculate silhouette. As we can clear see from this figures most of the data gives good fit within its clusters especially with k-means clustering, with Ward's clustering we see that third cluster is really artificial and we should stick with two clusters approach using this algorithm.

Our next goal was to check how accurate is our clustering i.e. are we getting same clusters from the both algorithms. In order to do so we used Matlab crosstab function, which returned chi-squared

IOP Conf. Series: Journal of Physics: Conf. Series **913** (2017) 012005 doi:10.1088/1742-6596/913/1/012005

value of 157.4737 for Pearson's chi-squared test of independence and corresponding p-value $5.0896*10^{-33}$, which means strong correlations between clusters obtained by both algorithms.

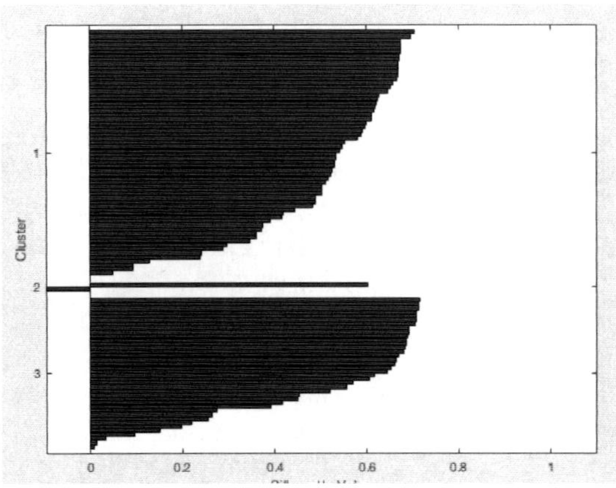

Figure 2. K-means clustering silhouette values

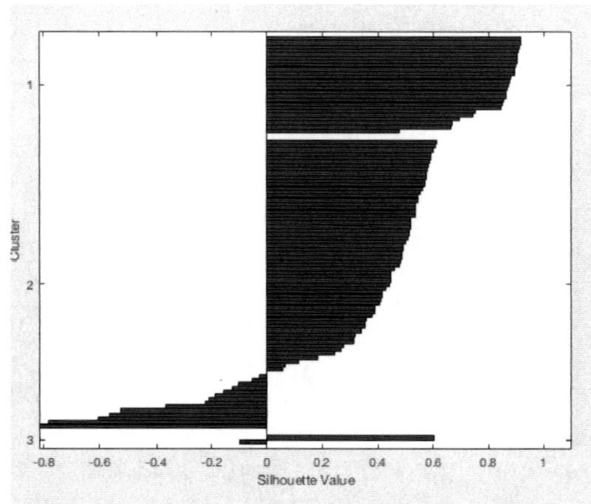

Figure 3. Ward clustering silhouette values

5. Conclusion

In this study we have clearly show that machine learning clustering techniques can be used to perform classification of the universities by the source of their income. The next logical step would be to cluster universities by their spending and overall P&L statements. This classification will allow us to quantitate classification of universities as teaching, research, entrepreneur, etc. which is important tool for government, corporations and general public alike in setting expectation and selecting universities to achieve different goals.

References

[1] Etzkowitz H 2003 Research groups as 'quasi-firms': the invention of the enterpreneurial universiry *Research Policy* **32** p 109–121

[2] Clark B 1998 *Creating enterpreneurial universities: organizational pathways of transformation* (New York)

[3] Etzkowitz H 1998 The norms of enterpreneurial science: cognitive effects of the new university – industry linkages *Research Policy* **27(8)** p 823–833

[4] Gibb A 2005 Towards the Enterpreneurial University: Enterpreneurship Education as a level for change [online] Avaliable from: http://ncee.org.uk/wp-content/uploads/2014/06/towards_the_entrepreneurial_university.pdf

[5] Guerrero M and Urbano D 2012 The development of an entrepreneurial unovercity *The Journal of Technology Transfed* **37(1)** p 43–74

[6] Gulbransden M and Slipersater S 2007 The third mission and the enterpreneurial university model, ed A. Bonaccorsi *Universities and Strategic Knowledge Creation*: *Specialization and performance In Europe* (Edward Elgar Publishing) p 112–144

[7] Kirdy D A 2006 Creating enterpreneurial universities in the UK: applying enterpreneurship theory to practice *Journal of Technology Transfer* **31(5)** p 599–603

[8] Montesinos P et al 2008 Thid Mission Ranking for World Class Universities: Beyond Teaching and Research *Higher Education in Eroupe* **33(2–3)** p 259–271

[9] Nelles J and Vorley T 2010 Constructing an Enterpreneurial Architecture: An Emergent Framework for Studing the Contemporary University beyond the Enterpreneurial Turn *Innovative Higher Education* **35(3)** p 161–176

[10] RETRACTED ARTICLE: University financial management concepts and mechanisms of innovation Sun (M.-C. 2010 Proceedings 2010) *2nd IEEE International Conference on Information and Financial Engineering* ICIFE 2010 5609412 p 517–521
[11] Wu C and Wang W 2011 University financial revenues and expenditures of the existing problems and countermeasures *2nd International Conference on Artificial Intelligence, Management Science and Electronic Commerce (AIMSEC 2011)* p 6895–6898
[12] Scaringella L and Chanaron J-J 2016 Grenoble – GIANT Territorial Innovation Models: Are investments in research infrastructures worthwhile? *Technological Forecasting and Social Change* **112** p 92–101
[13] Zonca P, Moravcova V, Maresova P and Kuca K 2015 Financing of a Public Universities in Czech Republic *Proceedings of the 25th International Business Information Management Association Conference – Innovation Vision 2020: From Regional Development Sustainability to Global Economic Growth (IBIMA 2015)* p 198–201
[14] Totska O L 2014 Structure of university revenues and expenditures in Canada and Ukraine: Comparative analysis *Actual Problems of Economics* **151(1)** p 225–232
[15] Bonollo E, Lazzini S and Zuccardi Merli M 2017 Accounting information system and organizational change: An analysis in "first mover" public Universities *Lecture Notes in Information Systems and Organisation* p 183–201
[16] Ponorîcă A G, Popa A F and Stănilă G O 2013 Comparative study of budgeting systems within various European universities *Quality – Access to Success 14* (SUPPL.2), p 53–65
[17] *Best Global Universities* Avaliable from: https://www.usnews.com/education/best-global-universities
[18] *U.S. Colleges and Universities-Utility Patent Grants, Calendar Years 1969–2012* Avaliable from: https://www.uspto.gov/web/offices/ac/ido/oeip/taf/univ/univ_toc.htm
[19] Kohonen T 1982 Self-organized formation of topologically correct feature aps *Biological Cybernetics* **43** p 59–69
[20] Kohonen T 2001 Self-Organizing Maps *Third, extended edition* (Springer)
[21] Ward J H, Jr 1963 Hierarchical Grouping to Optimize an Objective Function *Journal of the American Statistical Association* **58(301)** p 236–244
[22] MacQueen J B 1967 Some Methods for classification and Analysis of Multivariate Observations *Proceedings of 5th Berkeley Symposium on Mathematical Statistics and Probability* (University of California Press) p 281–297
[23] Steinhaus H 1957 Sur la division des corps matériels en parties *Bull. Acad. Polon. Sci.* (in French) **4(12)** p 801–804
[24] Lloyd S P 1957 Least squares quantization in PCM *IEEE Transactions on Information Theory* **28(2)** p 129–137
[25] Kriegel H-P, Schubert E and Zimek A 2016 The (black) art of runtime evaluation: Are we comparing algorithms or implementations? *Knowledge and Information Systems* **52(2)** p 341–378
[26] MacKay D 2003 *Information Theory, Inference and Learning Algorithms* (UK, Cambridge University Press)

IOP Conf. Series: Journal of Physics: Conf. Series **913** (2017) 012006 doi:10.1088/1742-6596/913/1/012006

Impact of the macroeconomic factors on university budgeting the US and Russia

Arina Bogomolova[1, 2], Igor Balk[2], Natalya Ivachenko[1] and Anatoly Temkin[3]

[1] Lomonosov Moscow State University, GSP-1, Leninskie Gory, Moscow, 119991, Russia

[2] Global innovation Labs LLC, 258 Harvard Str #352 Brookline MA 02446 USA

[3] Boston University Metropolitan College, 808 Commonwealth Avenue, Boston, MA, 02215, USA

e-mail:info@innovationlabs.net

Abstract. This paper discuses impact of macroeconomics factor on the university budgeting. Modern developments in the area of data science and machine learning made it possible to utilise automated techniques to address several problems of humankind ranging from genetic engineering and particle physics to sociology and economics. This paper is the first step to create a robust toolkit which will help universities sustain macroeconomic challenges utilising modern predictive analytics techniques.

1. Introduction

Technological progress during last half of the XX century significantly changed labor market all around the world. And therefore universities had to change as well to address to demand and economical realities. At first higher education, which depended heavily on the state, was perceived as a public good. But this belief was no longer valid [1]. Education was noticeably transformed under the pressure of numerous reforms. Economic stress increased the importance of the education system and the demand for quality improvement, which led to the need for greater funding. Higher education had to interact more closely with society and industry to increase the share of funding from the "market" and match the growing demand [2]. Now universities are even more closely related to the economy.

The main objectives of education are primarily knowledge maintenance and transfer. However, with the spread of the knowledge-based economy, universities have acquired new functions, such as technology transfer, building science and technology research parks, creature of incubators, providing consultancy services, etc. [3]. Therefore universities are becoming more and more dependent of the revenue from the sources outside of education and research and university budgeting process is becoming even more complex and sets higher standards for ROI expectations. In this way, budgeting – this is part of the economic cycle [4].

According to one of the definitions, budgeting is the decision on resources repartition among subunits of organization [5]. Frederick S. Hills and Thomas A. Mahoney tried to explain the process of budget allocation in universities. They came to the conclusion that universities as organizations behave according to the theory of coalition. Moreover, the method of allocating funding among subunits depends on the availability of resources. But the central power is more important in times of scarcity, which usually occurs during a crisis [6].

IOP Conf. Series: Journal of Physics: Conf. Series **913** (2017) 012006 doi:10.1088/1742-6596/913/1/012006

On the other hand, Joseph G. Fisher suggests using budgets for resource allocation that can eliminate budget slack. Moreover, the budget can be used for performance evaluation and this increases subordinate performance and more accurate budgets. Thus, management methods of budgeting, combining planning and control functions, can have a positive effect on the performance of the organization [7]. However, the authors made conclusions based on specific individual organizations and leave room for future research.

Another case study is an overview of budget allocations in one of the state universities in Malaysia. On the basis of mathematical calculations, the authors tried to find the effect of budget allocations for different faculties and subunits of the university. So they want to find a fresh approach to the resource allocation based on the effectiveness of the subunits [8].

Ingo Liefner analyzes various forms of allocation recourses and funding affect universities at the macro and micro levels. Private funding plays a more important role in Anglo-American high education system. But in each case, the characteristics of a particular university affect the success of the funding allocation scheme. There are also differences in short-term and long-term results, which must be taken into account [9]. Because of the effect of globalization, the world economy has a significant impact on the development of education [10]. The behavior of organizations can be even the opposite. For example, during the 2008 downturn, companies reduced their expenditure for R & D, as there were fewer resources and it was necessary to survive in the short term. However, some organizations have increased funding for research in the hope of future incomes [11].

However, there may be an inverse relationship between education policy and funding. Alexander L. Darling and Martin D. England studied the financial formula in Ontario and assessed the effectiveness of budget allocation as a public policy tool [12]. The economy and the resulting indicators of education have a mutual impact. The crisis is an unpredictable and dangerous fluctuation of the economic structure, which must affect the education system.

The goal of this paper is to study which factors impact university budgeting in the current economical reality. This is a first step in the research project, which is aiming to use machine learning techniques to create automated university budgeting tool helping institutions of higher education be better prepared to macroeconomic fluctuations and become more sustainable in there development.

2. Methodology and data

The study of financial flows is a convenient tool for analyzing the activities of the organization as well as important mechanism in organizational health forecasting. The advantage of this type of analysis is that it can produce dynamical picture from both qualitative and quantitative points of view. Most of financial data is provided in uniform and easy to normalize format, but even with this in consideration a number of simplifications should be taken in the course of the study.

In this paper we considered two major indicators describing of the US economy: GDP in comparable prices of 2015 in US dollars and the Dow Jones index average for the year. The Dow Jones index reflects the dynamics of entrepreneurial activity. It is more sensitive to macroeconomic fluctuations than GDP. Similar indicators were taken for Russia: GDP in comparable prices in 2015 in Russian rubles and the RTS index average for the year. RTS index is the main indicator of the dynamics of market shares of leading companies in Russia. The above-mentioned indicators provide an opportunity to evaluate the health and dynamics of the economy in the country at the macroeconomic level.

Against the backdrop of the dynamics of macroeconomic indicators, we consider the system of research and development funding. For the US, we used the data provided by American Association for the Advancement of Science (AAAS) [13]. There are a number of federal R&D data series, including research budgets by agency, character and discipline. Most data come from annual AAAS R&D budget reports. Additionally we used data from the National Science Foundation's National Center for Science and Engineering Statistics [14], and data provided by the White House Office of Management and Budget (OMB) [15]. Total R&D funding and funding by performer among which federal, industry, university, federally funded R&D centers and nonprofits were taken in 2015 US dollars.

For Russia, we used data from Ministry of Education and Science of the Russian Federation summarized in the report that was prepared by the Republican Research Scientific-consulting Center of Expertise [16]. The available indicators for the study are total funding and funding by performer such as federal, industry, university and nonprofits. The data were brought to a comparable way by adjusting for the official value of inflation to the rubles 2015.

Further the university budgets were considered in more details. US data on funding sources for university R&D are collected by the National Science Foundation's National Center for Science and Engineering Statistics [17]. According to this agency we can name such sources of R&D funding in universities as federal, state and local, universities, industry and other. The values were translated into the prices in 2015 US dollar using the inflation factor. In Russia according to the data from above-mentioned report universities receive funding from higher education institutions, federal, industry and foreign sources, off-budget funds, own funds and funds of private non-profit organizations. The values are also translated into the prices in 2015 Russian rubles using the inflation factor.

For comparison between US and Russia we used the Purchasing power parities (PPPs) index. Purchasing power parities are the rates of currency conversion that equalize the purchasing power of different currencies by eliminating the differences in price levels between countries and we used World Bank Data to normalize the data [18].

To study a data correlation we used simple Paerson correlation coefficient, which is a measure of linear correlation between two variables X and Y. [19] It can be described by simple formula

$$r = \frac{\sum_{i=1}^{n}(x_i-\bar{x})(y_i-\bar{y})}{\sqrt{\sum_{i=1}^{n}(x_i-\bar{x})^2}\sqrt{\sum_{i=1}^{n}(y_i-\bar{y})^2}},$$

where n is the number of samples, x_i, y_i are the samples with respect to the index i, $\bar{x}$ and $\bar{y}$ are mean values of the sample. And the goal was to find a correlation between both overall university R&D expenditures and macroeconomic indicators as well as such correlation between macroeconomics indicators and various financial sources for R&D activities. This step is necessary to create accurate model, which will forecast optimal university R&D budget based on macroeconomics modeling.

3. Data analysis

First we will consider dynamics of total R&D spending's, depending on the state of the economy in the country. As we have mentioned before GDP is used as a major indicator of the economic grows. As we can see from the Figures 1 and 2 total R&D spending almost tracking GDP in both Russia and the US (corresponding correlation coefficients are 0.9412 for the US and 0.9779 for Russia). If we consider in more detail each series of values, then it can be seen that the largest deviations between R&D funding and GDP correspond to the economy up and down turns. For instance, during the recession GDP had a local minimum, but at the same the total R&D spending's have local maximum and vice versa. This can be seen in Figure 1 in 2001 and 2008, and in Figure 2 in 2003 and 2009, although to a lesser extent. It should be noted that although extremes are close to each other, they are not necessary in the exactly same year. Thus, we can conclude that the R&D spending's are growing accordingly with the growth rate of the economy. The most notable deviation is most likely caused by the recessions of the 2000s and 2008.

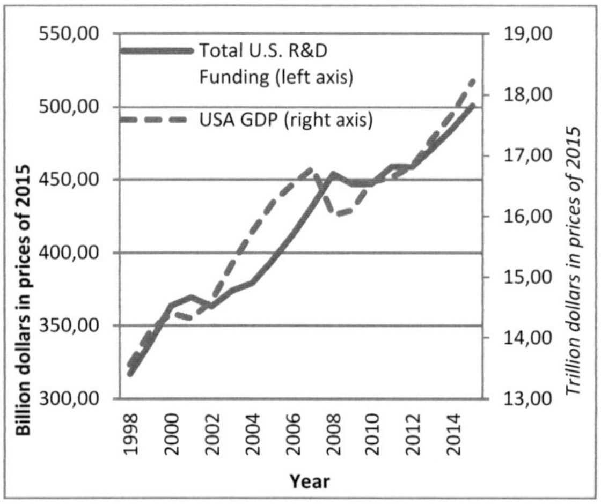

Figure 1. TotalR&D spending's vs GDP in the US (in 2015 US dollars)

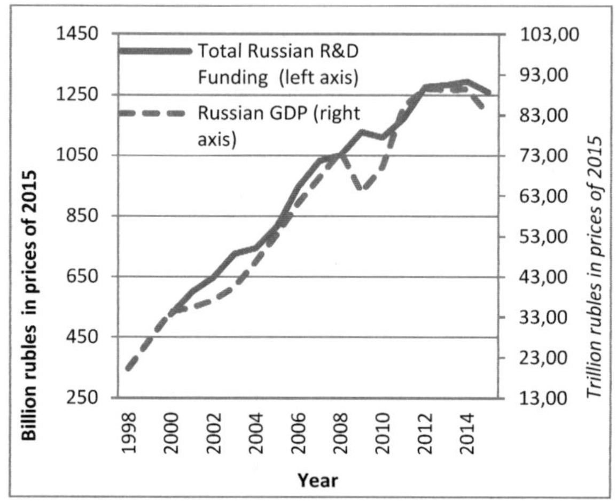

Figure 2. Total R&D spending's vs GDP in Russia (in 2015 Russian rubles)

Another important indicators of the health of economy are stock market indexes as The Dow Jones for the US RTS for Russia. Stock markets are much more volatile than GDP and thus could be much better macroeconomic indicators. In this scenario we still can see significant correlation between macroeconomic factors and R&D spending's (Figures 3 and 4, corresponding correlation coefficients are 0.7691 for the US and 0.7423 for Russia).

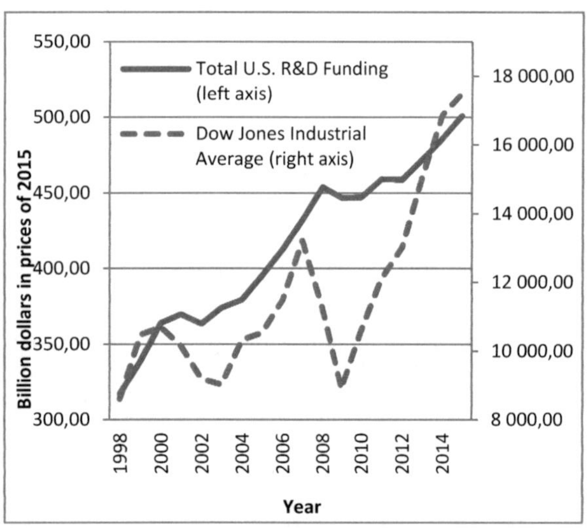

Figure 3. R&D spending's vs Dow Jones in the US

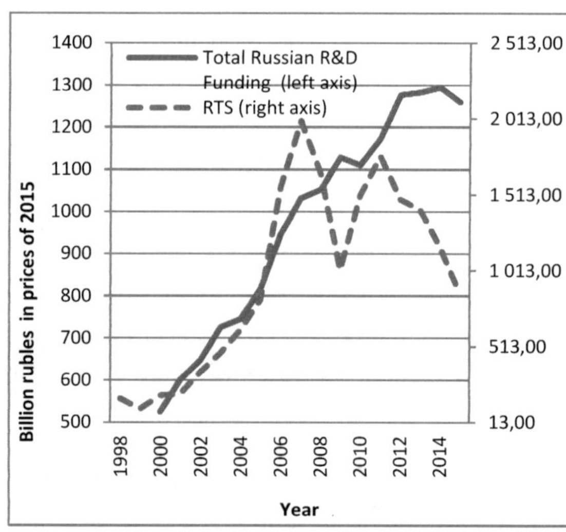

Figure 4. R&D spending's vs RTS in Russia

Nevertheless we can see significant difference in dynamics of R&D funding between the US and Russia if we use PPP index in 2015 US dollars. As we can see from the Figure 5 the US R&D funding grew 1.66 times from 1998 to 2014 while Russian R&D spending is almost flat over the same period of time.

Now let's look in more detail what are the reasons for fluctuations of R&D spending's. First, we will compare the most active performer in R&D (the Industry) with Universities contribution.

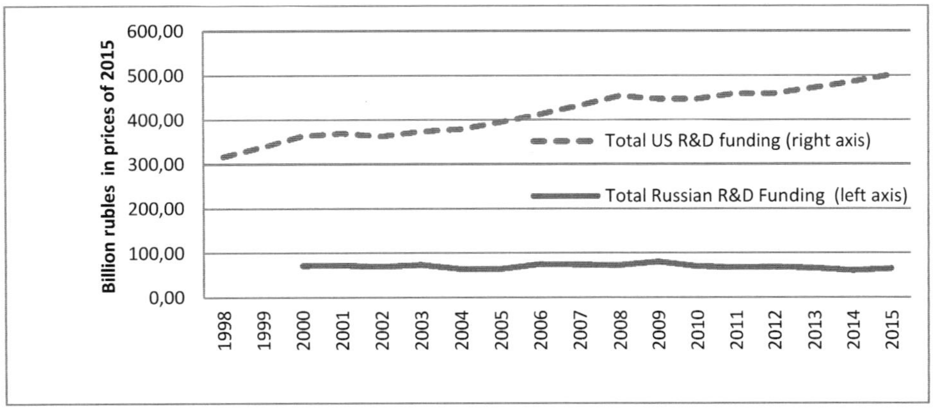

Figure 5. R&D spending's US vs Russia (in 2015 US dollars)

As we can see from the Figure 6 for US, universities operate in opposition to the industry. The university does not always react quickly to changes and therefore the extremes do not coincide but the main trend can be seen very clear: universities compensate for the luck of industrial R&D spending's during economy downturns. (Correlation coefficient 0.8571) In Russia the situation is completely different (Figure 7): universities behave the same way as the industry, strengthening the total fluctuations. (Correlation coefficient 0.94) Strong values of the correlation coefficient also support this theory.

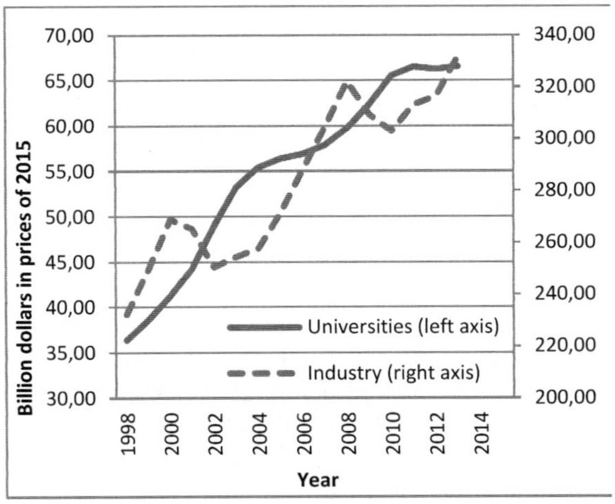

Figure6.R&D spending's in Industry vs Universities in the US (normalized)

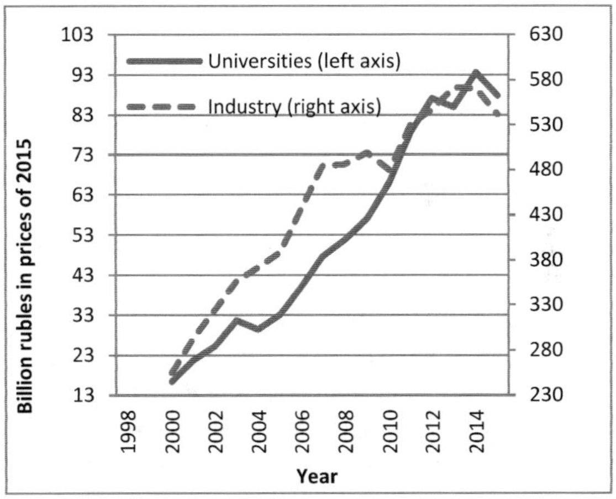

Figure 7. R&D spending's in Industry vs Universities in Russia (normalized)

As can be seen from the Figure 8 and 9 both Russian and US government are trying to mitigate negative impact of macroeconomic downturns on the university R&D activities. It worth to mention that both in the US as well as in Russia federal government funding remain main source of the university R&D budget. Such funding accounts to about 70% of total university R&D spending emphasizing huge role of the government in the modern science.

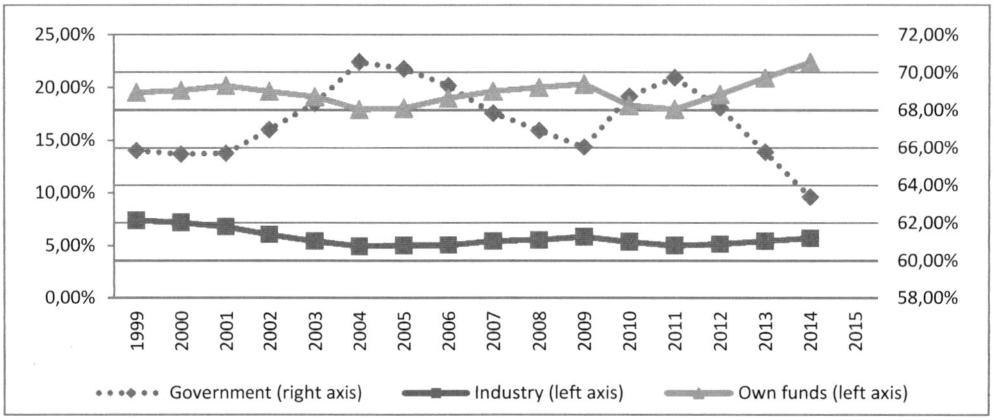

Fig 8. Sources of R&D financing in the US universities (normalized)

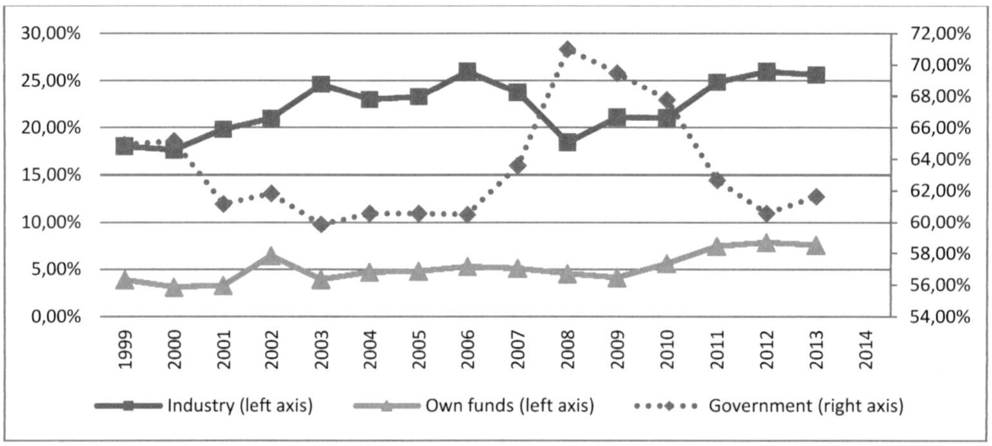

Figure 9. Sources of R&D financing in the Russian universities (normalized)

4. Conclusion and future work

As we have discussed in the previous chapters we can see a strong correlation between macroeconomic factors and university R&D spending. It worth to note that government and industry financing of R&D activities show opposite behaviors in respect to macroeconomic downturns. Therefore financial diversification could be a key to sustainable financial model of university budgeting.The logical next step would be study of other income sources for the universities such as tuition, endowment, technology transfer etc. and development of the optimal P&L model

References

[1] Jessop Bob 2017 Varieties of academic capitalism and entrepreneurial universities *Higher Education* **73(6)** p 853–870
[2] Hills F S and Mahoney T A 1978 University Budgets and Organizational Decision Making *Administrative Science Quarterly* **23(3)** p 454–465
[3] Kogan M, Bauer M, Bleiklie I and Henkel M 2000 Transforming Higher Education *Higher education dynamics* **13**, ed. Peter Maassen (London: Jessica Kingsley Publishers)
[4] Fisher J G, Maines L A, Peffer S A and Sprinkle G B 2002 Using Budgets for Performance Evaluation: Effects of Resource Allocation and Horizontal Information Asymmetry on Budget Proposals, Budget Slack, and Performance *The Accounting Review* **77(4)** p 847–865
[5] Ingo Liefner 2003 Funding, Resource Allocation and Performance in Higher Education Systems *Higher Education* **46(4)** p 469–489

[6] Lepori B, Usher J and Montauti M 2013 Budgetary allocation and organizational characteristics of higher education institutions: A review of existing studies and a framework for future research *Higher Education* **65(1)** p 59–78

[7] Enders Jürgen 2008 Creating public-private dynamics in higher education funding. A discussion of three options *Public-Private Dynamics in Higher Education* ed Ben Jongbloed p 113–138

[8] Rossidah Wan Abdul Aziz, Adibah Shuib, Wan Noor Hayatie Wan Abdul Aziz, Norngainy Mohd Tawil and Abdul Halim Mohd Nawawi 2013 Pareto Analysis on Budget Allocation for Different Categories of Faculties in Higher Education Institution *6th International Conf. on University Learning and Teaching (Procedia – Social and Behavioral Sciences)* **90** p 686–694

[9] Cincera M, Cozza C, Tübke A and Voigt P 2012 Doing R&D or Not (in a Crisis), That Is the Question... *European Planning Studies* **20** p 1525–1547

[10] Julia Resnik 2008 Introducing a Neo-Weberian Perspective in the Study of Globalisation and Education: Structural Reforms of the Education Systems in France and Israel after the Second World War *Oxford Review of Education* **34(4)** p 385–402

[11] Andrew Wall, Robert Frost, Ryan Smith and Richard Keeling 2008 Examining a Higher Education Funding Formula in a Time of Shifting Currents: Kentucky's Benchmark Approach *Journal of Education Finance* **33(3)** p 221–237

[12] Darling A L, England M D, Lang D W and Lopers-Sweetman R 1989 Autonomy and Control: A University Funding Formula as an Instrument of Public Policy *Higher Education* **18(5)** p 559–583

[13] Historical Trends in Federal R&D *American Association for the Advancement of Science* (AAAS) Avaliable from: https://www.aaas.org/page/historical-trends-federal-rd

[14] Federal Funds for R&D *National Science Foundation's National Center for Science and Engineering Statistics* Avaliable from: https://www.nsf.gov/statistics/fedfunds/

[15] *White House Office of Management and Budget* (OMB) Avaliable from: https://www.whitehouse.gov/

[16] *R&D spending and its sources* 2016 Avaliable from: http://csrs.ru/archive/stat_2016_finance/

[17] R&D at Colleges and Universities *American Association for the Advancement of Science* (AAAS) Avaliable from: https://www.aaas.org/page/rd-colleges-and-universities

[18] PPP conversion factor GDP (LCU per international $) *World Bank Data* Avaliable from: http://data.worldbank.org/indicator/PA.NUS.PPP?end=2015&locations=RU&start=1998&view=chart

[19] Karl Pearson 1895 Note on Regression and Inheritance in the Case of Two Parents *Proceedings of the Royal Society of London* **58** p 240–242

Analysis of user activities on popular medical forums

M V Kamalov[1, 2], V Y Dobrynin[1], Y E Balykina[1] and R S Martynov[1, 2]

[1] Applied Math and Control Processes Department, Saint Petersburg State University, Saint Petersburg 198504, Russia
[2] EPAM systems, Saint Petersburg 196084, Russia

E-mail: mkamalovv@gmail.com

Abstract. The paper is devoted to detailed investigation of users' behavior and level of expertise on online medical forums. Two popular forums were analyzed in terms of presence of experts who answer health related questions and participate in discussions. This study provides insight into the quality of medical information that one can get from the web resources, and also illustrates relationship between approved medical experts and popular authors of the considered forums. During experiments several machine learning and natural language processing methods were evaluated against to available web content to get further understanding of structure and distribution of information about medicine available online nowadays. As a result of this study the hypothesis of existing correlation between approved medical experts and popular authors has been rejected.

1. Introduction

Evaluating quality of available information was always a hard to solve problem, especially nowadays, when the data grows with incredible velocity. Content, published on the web by users, always requires verification, especially if it considers such a field as healthcare. Experts in this area could provide an incredibly useful information, but there are also people with lower level of expertise, whose incorrect interpretation or formulation of some concept could lead to a real damage to someone's health. Moreover, an instrument which would allow to divide experts from others could be useful for research purposes, such as finding special patterns in language, sentiment analysis, etc.

There were several approaches to similar problems. Researches around the world studied users' behavior online in terms of social grouping [1], community-based methods [2] and clicks and ratings in search engines and web stores. Another subject was analysis of interconnections between unstructured data on the web (discussion forums and well-known sources of information) [3]. Also, there are studies in health-related information found on the Internet: log [4] and search queries analysis aimed at evaluating users' preferences and expertise [5]. By the way existing researches which focused on analysing abstracts of medical studies from the perspective of evidence based medicine [6, 7]. This article focused on quality evaluation of conducted medical studies by analysing text abstracts. But most of them provides either too general information without deep insights to specific semantic areas or operates on highly structured data (e.g. NCBI PubMed database). This paper presents deep analysis of medicine-related content based on unstructured information from online forums.

IOP Conf. Series: Journal of Physics: Conf. Series **913** (2017) 012007 doi:10.1088/1742-6596/913/1/012007

2. Data description

We used content from two popular health-related forums: doctorslounge.com and patient.info. We extracted the data using web crawling instruments, saving posts text, information about authors and hierarchical structure (forum categories) in terms of each forum labeling system. The overall data is shown in Table 1.

Table 1. General characteristics of data

	doctorslounge.com	patient.info
Total amount of data (number of posts)	143921	1476990
Number of top level categories	22	31
Number of second level categories	95	325
Number of specific topics ("branches" of discussion)	25843	157754

Further exploration was based on the fact that the doctorslounge.com forum has labels for experts in medicine area – it is a highly moderated forum with authentication procedure, while patient.info provides only information about "popularity" of users. That means, that in patients.info forum there is a system of levels of expertise (there are six of them), which is based on users' activity and social approval. Based on that data we formulated the following research questions:

RQ1: Is there a correlation between level of user expertise from patiens.info and medical doctors from doctorslounge.com?

RQ3: Does the forum internal structure affect expert distribution and the amount of produced content?

To answer these questions, we first looked at the existing distribution of users with respect to their labels (figure 1, figure 2). In the doctorslounge.com data there were more specific labels for experts (such as "Paramedic" and "Psychiatrist"), so we merged them into one label "Medical Doctor", which was already presented. In the patiens.info data users were divided into 6 levels which meant rank of expertise and popularity of author on this forum. Each level has its own identification labels: level 1 – beginner, level 2 – Contributor, level 3 – Regular, level 4 – Accomplished, level 5 – Guru, level 6 – Oracle.

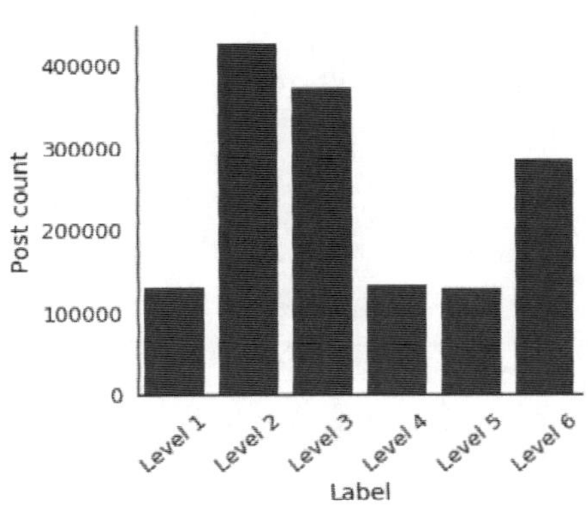

Figure 1. Users' post count by label on doctorslounge.com data

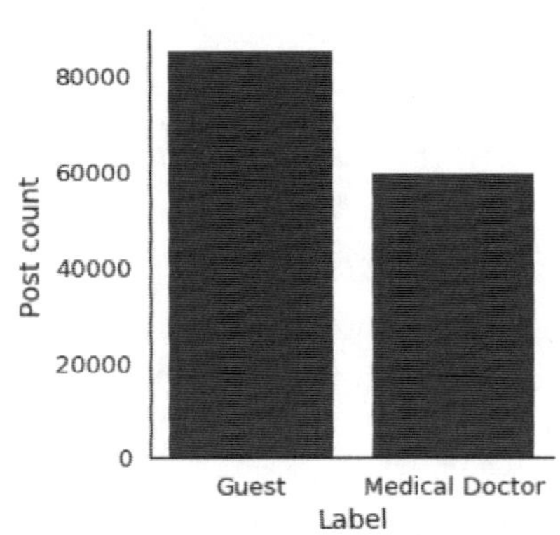

Figure 2. Users' post count by label on patient.info data

Based on this data we focused on the main hypothesis about difference between groups of users was that they language, i.e. words and collocations, would differ a lot. To test this, we first built a classification model based on doctorslounge.com data.

The whole text corpus (posts, titles) was preprocessed for evaluation: all words were transferred to lower case, stemmed using Porter stemmer, and all stop words were removed. Then a TF-IDF representation of processed words and user labels as target values was used as an input to classification algorithm (considering sparsity of data we used logistic regression). Quality of the model was checked on 10-fold cross validation (figure 3) in terms of F1-score and AUC-ROC – two well-known metrics used in classification tasks. As one can see, it fitted existing data.

Then the predictions were made on the patient.info data, preprocessed in the same way. After that we had labels for each posts determining whether it was written by an expert (or "Medical doctor" in terms of doctorslounge.com labeling) or not.

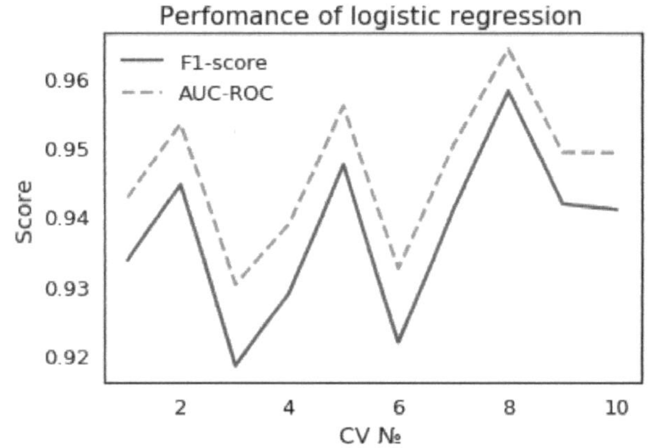

Figure 3. Performance metrics of classification model

3. Labels relationship

To answer the RQ1, we studied the distribution of newly labeled experts over existing level system (figure 4). As one can see, the distribution is almost uniform in a sense that each level contains almost equal proportion of experts (numbers inside bars). That means that there is only a small correlation between real expertise and social-based leveling in the data.

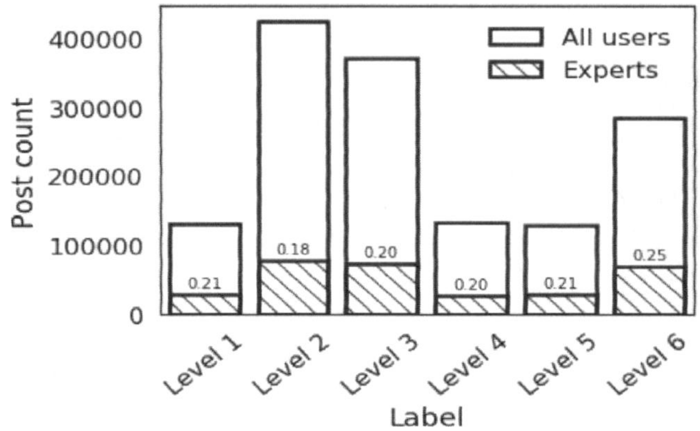

Figure 4. Experts' distribution over levels in patient.info data

IOP Conf. Series: Journal of Physics: Conf. Series **913** (2017) 012007 doi:10.1088/1742-6596/913/1/012007

Other approach was to compare the vocabulary of experts and less experienced users in order to find out, whether there is any strong dependency. To do that we built a distribution of words from each class over user labels. These distributions allow us to measure the distance between them in terms of Jensen-Shannon divergence, which corresponds to similarity between vocabularies and overall language use in classes. Jensen-Shannon divergence is defined as follows:

$$\mathrm{JSD}(\mathrm{P} \,||\, \mathrm{Q}) = 0.5 * \mathrm{D}(\mathrm{P} \,||\, \mathrm{M}) + 0.5 * \mathrm{D}(\mathrm{Q} \,||\, \mathrm{M}),$$

where $\mathrm{M} = 0.5 * (\mathrm{P} + \mathrm{Q})$ and $\mathrm{D}(\mathrm{PM}) = \sum_{\mathrm{i}} \mathrm{P(i)} \log \frac{\mathrm{P(i)}}{\mathrm{Q(i)}}$ – the Kullback-Leibler divergence.

We used a distribution of words over "Medical Doctor" label (P_{MD}) based on doctorslounge.com data as a base distribution and measured distance from it to all others. Results could be found in Table 2. We also calculated the divergence value between P_{MD} and distributions of words of users that our model labeled as experts as an additional sanity check for classification algorithm (if our classification is correct, the divergence value between those two distributions must be relatively small).

Table 2. General characteristics of data

Label corresponding to distribution P_l	$JSD(P_l \,||\, P_{MD})$
Guest	0.1109
Level 1	0.1362
Level 2	0.1354
Level 3	0.1397
Level 4	0.1408
Level 5	0.1419
Level 6	0.1251
Predicted expert	0.0956

The values are very close to each other for the level-based labeling system (except for Level 6), so we can not rely on that if we want to divide users by their expertise. The values also suggest that there is no strong correlation between vocabularies of users at each level and experts.

To further investigate possible dependency we decided to look at experts distributions over existing top 5 patiens.info forum categories which contain more classified users as medical doctors than others (figure 5).

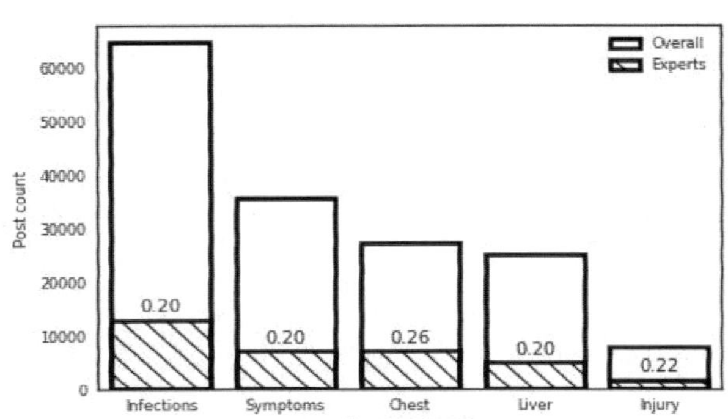

Figure 5. Medical Doctors over patiens.info category

IOP Conf. Series: Journal of Physics: Conf. Series **913** (2017) 012007 doi:10.1088/1742-6596/913/1/012007

Based on the results presented in figure 5 we have not found a category which contains more classified medical doctors than others. It means that distribution of medical doctors over forum category looks like uniform. Thus, these results can not completely answer the RQ2 question. However more detailed representation of classified medical doctors over levels of expertise is presented in figure 6 and 7 for categories "infections" and "injury". This representation allow us to notice that the level 6 contain more doctors than others.

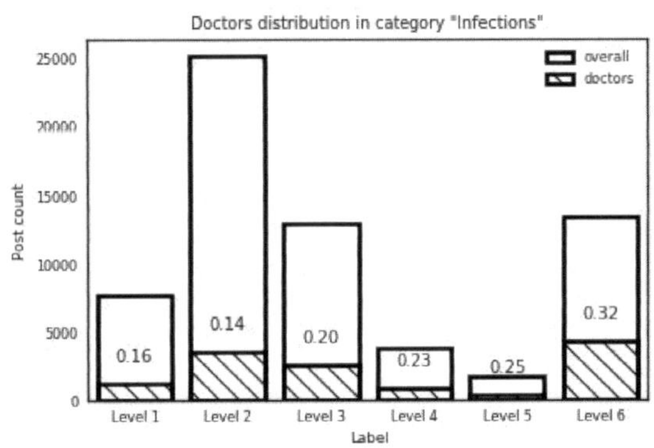

Figure 6. Medical doctors present over levels in "Infections" category

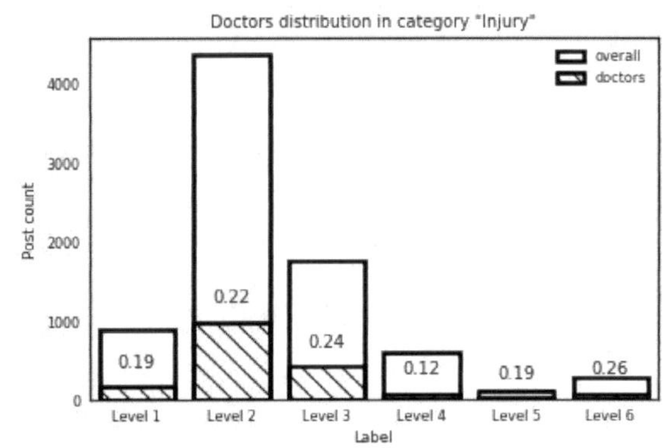

Figure 7. Medical doctors present over levels in "Injury" category

However forum category does not completely answer at the questions RQ1, RQ2. Those result led us to use topic modelling approach.

4. Approach based on topic modelling

The topic modelling approach available to extract topics based on the post semantics. These topics describe medical areas in health-related forums. For this approach we applied Latent Dirichlet Allocation (LDA) algorithm which has been approved as the best solver for topic modelling task.

At the first step of topic modelling we preprocessed data for the LDA training and testing set. The pre-processing consist of tokenization, cast to lower case, stemming by Porters' stemmer, and removing stop words and words which contain less than 3 letters and higher than 10 letters. After that we divided posts for training and testing sets: – 600000 posts for train; – 200000 posts for test. Based on the grid search techniques and using perplexity as an evaluation measure we started the process of topic number selection. The result of grid search is presented in figure 8.

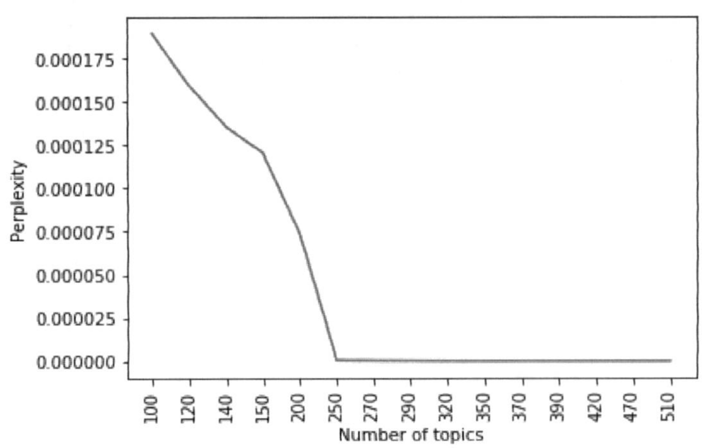

Figure 8. Perplexity over topics

Based on the obtained results we can see that optimal amount of topics equals 270. Thus, we get representation of doctors over most popular topics. Under the most popular topics we mean topics which contain a lot of user posts. From this point we got doctor distribution presented in figure 9. All topics in figure 9 is represented as a numerical index (id) which has been obtained from gensim lda[1]. This topics has been selected as containing more Medical Doctors than others.

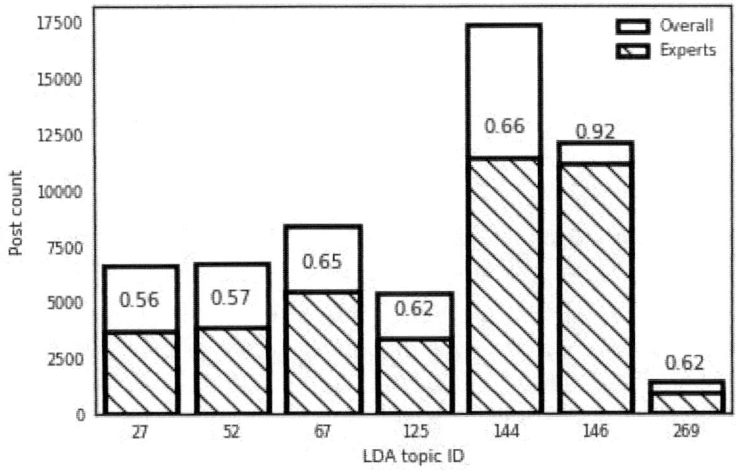

Figure 9. Doctor distribution over topics

Each topic index gets its own representation in terms of words distributions. In table 3 we get definition for each topic as words distribution.

Table 3 Most probably 5 words for topics

Topic index	Most probably words
27	examine; awar; stool; symptom; asthma
52	infect; sinu; chronic; color; treatment
67	disea; cell; allergi; histori; lead
125	women; abnorm; vagin; determin; symptom
144	evalu; bone; best; pain; wish
146	surgeon; opinion; second; schedul; pictur
269	oral; lesion; amanda; blockag; transmit

Additionally we prepare more detailed representation of classified medical doctors over levels of expertise is presented in figure 10, 11 for topics 144 and 146.

[1] https://radimrehurek.com/gensim/index.html

IOP Conf. Series: Journal of Physics: Conf. Series **913** (2017) 012007 doi:10.1088/1742-6596/913/1/012007

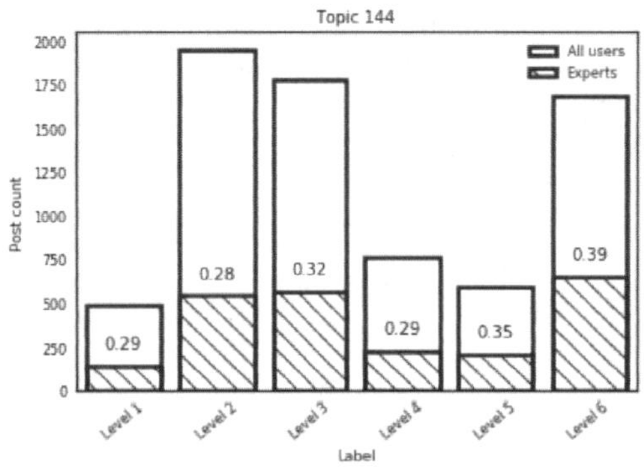

Figure 10. Medical doctors present over levels in topic 144

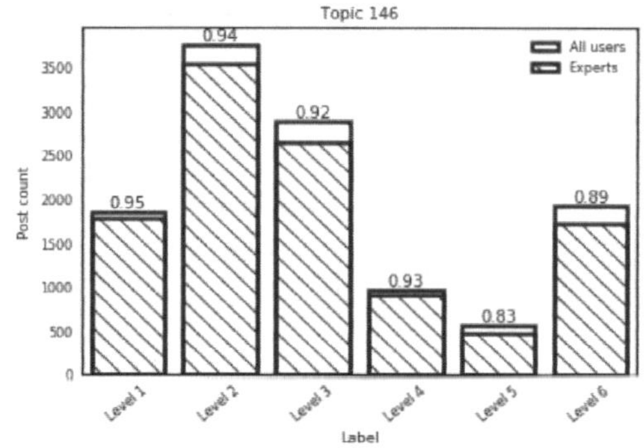

Figure 11. Medical doctors present over levels in topic 146

This results allow us to notice that the level 6 contain more doctors than others and topic modelling approach can extract semantic categories which has highest probability of a post written by a Medical Doctors. In our case we found topic 146 (surgeon; opinion; second; schedul; pictur) which get more the 90 present users classified as Medicals Doctors.

5. Conclusion

In this article, we presented in-depth analysis of health-related information found on two major forums: doctorslounge.com and patient.info. We carefully studied the level of expertise of users on both of these forums. As the result, we have strong evidence that there is no significant correlation between real expertise and the level of "popularity", which is based on analysis of forums internal hierarchical structure, experts vocabulary and language use.

We also proposed an approach to identify experts based on the topic modelling. Analysis of topic distribution allowed us to find semantic categories which has the highest probability of a post written by an experts.

For further research it is planned to append in topic modelling information from Medical subject headings (MeSH). We suppose that this structure could increase performance of extracting topics from post and reject uninformative topics. Additionally we start looking methods for extracting special behavioral patterns in content provided by medical experts.

References

[1] Xiaolin Shi, Jun Zhu, Rui Cai and Lei Zhang 2009 User grouping behavior in online forums *Proc. of the 15th ACM SIGKDD Int. Conf. on Knowledge discovery and data mining (KDD 2009)*

[2] Rowe M, Fernandez M and Alani H 2013 Modelling and analysis of user behaviour in online communities *STCSN E-letter* **1(2)**

[3] Pendry L and Salvatore J 2015 Individual and social benefits of online discussion forums *Computers in human behavior* p 211–220

[4] Dogan R, Murray G, Névéol A, and Lu Z 2009 Understanding PubMed® user search behavior through log analysis *Databse (Oxford)*

[5] Illhoi Y and Mosa A 2015 Analysis of PubMed user sessions using a full-day PubMed query log: a comparison of expirienced and nonexpirienced PubMed users *JMIR Med Inform*

[6] Kolbin A, Kamalov M, Dobrynin V, Balykina Y and Verbitskaya E 2016 Development of a Search Engine For The Medline Database With Search Results Ranking From the Perspective Of Evidence-Based Medicine *Value in Health* **19(7)** A369

[7] Kamalov M et al 2015 Improving data retrieval quality: evidence based medicine perspective *International Journal of Risk & Safety in Medicine* **27 s1** p 106–107

Fractal techniques as alternative of Fourier transform for fast search of periodic events in time series

A Prosvetov

Space Research Institute, Russian Academy of Science
Cleverdata

e-mail: prosvetov@gmail.com

Abstract. In our day Fourier analysis is commonly used in varies fields, that require information extraction form time series data. However, the Fourier technique has own pros and cons and we can found, that more contemporary approaches can give significant improvements in some cases. We take the example of astrophysical problem and try to resolve it with Fourier analysis in comparison with Fractal analysis. From our results one can find, that in cases of noise extraction Fractal techniques are more efficient and decrease the time of analysis of time series. The proposed approaches can be applied on other fields, to analyze the data sets with dramatically increased volume.

1. Introduction

In Astrophysics exist a problem of quasi-periodic oscillations in power spectra of X-ray light curves of black holes: the current physical models can't gives a complete, comprehensive picture of their formation, while the QPO generation mechanism seems to remain completely unexplained. The constructed power spectra usually consist of one or more band-limited ("red") noise components and, occasionally, quasi-periodic oscillation (QPO) peaks at low (0.1–1 Hz) and/or high (~10 Hz) frequencies (figure 1). Being the second-order statistical moments, the power spectra inevitably carry only limited information about the nonlinear variability mechanisms compared to the original light curves. Band-limited noise in the power spectrum is one of the signatures of not just a random but chaotic behavior of a physical system, i.e., the existence of fractal properties: this means a scale invariance, a self-similarity on various time scales. The chaotic variability, moderate on a time scale of tens of seconds, is seen to be repeated and amplified manyfold when passing to seconds and fractions of a second. To determine the degree of scale invariance of a light curve more rigorously (quantitatively), its fractal or Hausdorf dimension can be calculated. In particular, the degree of nonlinearity of the equations describing the physical process can be judged from the fractal dimension. Also the Hausdorf dimension of time series is connected to the noise power spectrum, and that fact gives a opportunity to extract information about noise spectrum without Fourier transform. The connection between fractal dimension of light curve of black holes and QPO is studied in our previous work [1]. The previous main result was that a correlation between the fractal dimension of the light curves and the frequency of the QPO peak has been revealed.

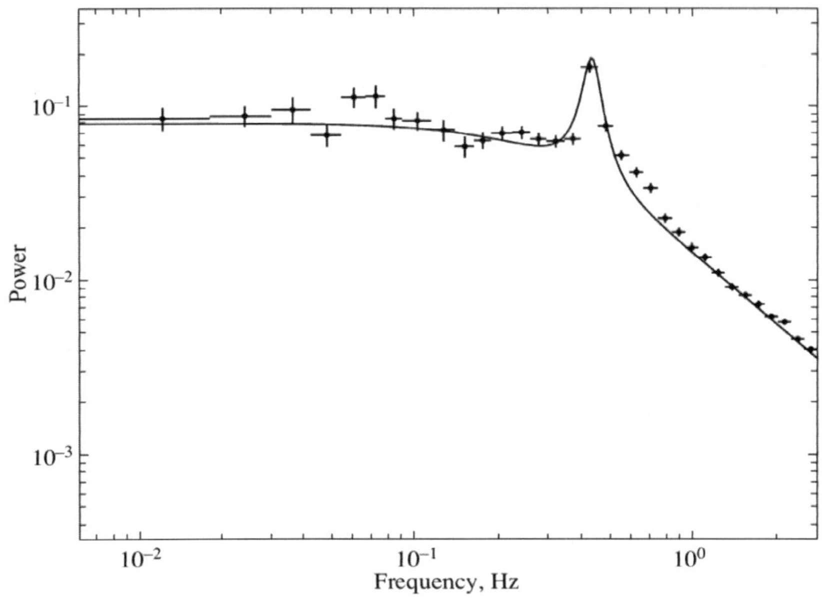

Figure 1. Power spectrum of GX 339-4 from its PCA/RXTE observations on February 16, 2007, in the energy range 2.9– 36.0 keV. The data were fitted by King's model with a line that had a Lorentz profile. The power is given in percentage normalization after subtraction of the Poisson noise. Powerful "red" (band-limited) noise with the QPO peak at 0.32 Hz is seen

In current paper we propose to use fractal techniques for fast search of observations with QPO in big arrays of observational data.

2. Data and instruments

For our study we used history observations of black holes (GX 339-4, Cyg X-1, Swift J1729) obtained with the following X-ray observatories: Integral, Swift, RXTE. The data were processed with the NASA/HEASARC software package. For our analysis, we used data in a wide energy range, 2.9–36 keV. The size of the time bin Δt was always taken to be 0.01 s.

We used the following algorithm (usually called R/S analysis) to determine D. We divided the time series being investigated into intervals and calculated the following in each of them: (1) the range R equal to the difference between the maximum and minimum of the function of the accumulated deviation from the mean in the interval, (2) the accumulated standard (root mean square) deviation from the mean S, and (3) the number of points in the interval N. For all intervals with the same N the value R/S is averaged. For all intervals with different N, we constructed the dependence of ln(R/S) on ln(N/2) and fitted the derived points of the dependence by a straight line using the least-squares method. The slope of the straight line specifies the Hurst exponent H. The Hurst exponent allows the fractal dimension of the time series to be determined, which is $D = 2 - H$ and reflects the scale invariance of the time series [2].

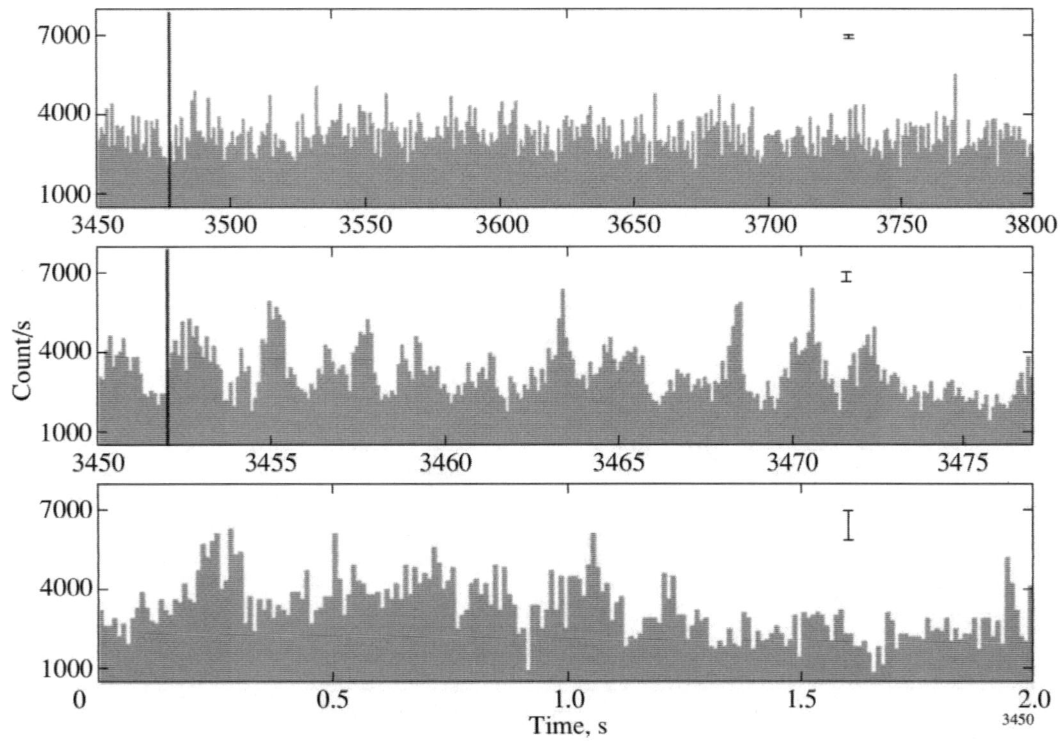

Figure 2. PCA/RXTE light curve from GX 339-4 in the energy range 2.9–36 keV. The typical measurement errors are shown on the right. The parts of the light curves separated by the vertical line on the upper panels are given on the lower panels with a better angular resolution

For additional interpretation we also performed the following operations with the data, by analogy with the approach to analyzing time series called Max-Spectrum [3]. The distribution of extrema depending on the time scale in the signal is also analyzed in the mentioned Max-Spectrum approach. To completely smooth out the correlations in the signal without changing its amplitude distribution, a random sample is additionally constructed from the original sample, with the new signal being obtained from the randomly mixed old signal. In our case, the ln(R/S) − ln(N/2) diagram of the new time series will differ from the original one. In the Max-Spectrum approach, $y = \alpha x + const$ for the mixed signal and $y = \alpha x + a\ log_2(\theta) + const$ for the original signal, which allows the parameter θ to be determined. The original of this statistical signal from the Max-Spectrum analysis is called an extremal index and it has been studied in detail by Leadbetter et al. (1983). The informative interpretation of the extremal index is that θ^{-1} is equal to the mean size of the cluster of extremes in the time interval being investigated [4].

3. Results

The form of typical light curve is represented on figure 2 and is highly jagged on a scale of 1 s (top) and on scales of 0.1 s (middle) and 0.01 s (bottom). There is a self-similarity of the light curve on various scales suggesting that the light curve can possess the property of fractality. The comparison of Hust exponent for two observations is present on the figure 3. One of the observation has red noise without registered QPO, the other has QPO in power spectrum.

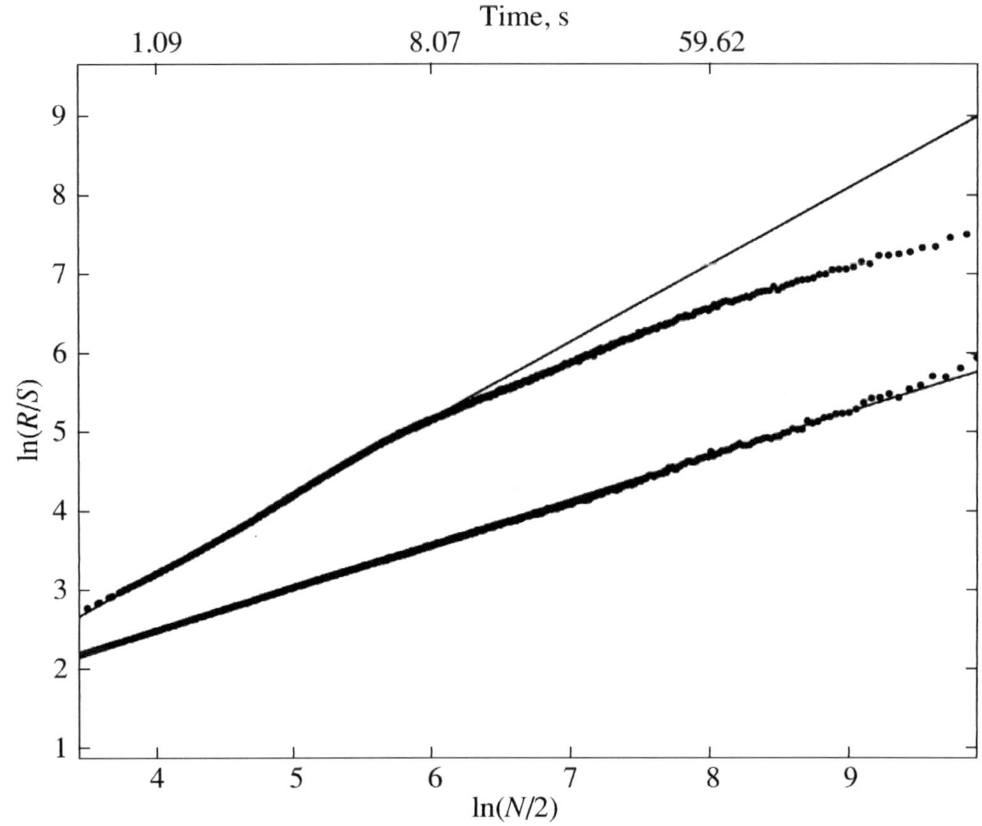

Figure 3. ln(R/S) − ln(N/2) diagrams for GX 339-4: at the top with QPOs (D = 1.02 for the scale ln(N/2) < 6); at the bottom without QPOs (D = 1.45). Along the axes: R is the range equal to the difference between the maximum and minimum of the function of the accumulated deviation from the mean in the interval; S is the accumulated standard deviation from the mean; and N is the number of points in the interval

One can find, the slope change on time scale corresponded to QPO frequency. The slope change can be understood the following way:

1) for time scales larger than period of QPO oscillations and Fourier analysis and R/S analysis can find only red noise in time series;

2) for time scale smaller than period of QPO oscillations Fourier analysis can observe only red noise again, but R/S analysis is highly dependent on peaks presence in time series. Due to periodicity of peaks, the majority of intervals for R/S calculations on the selected time scale have QPO oscillation peaks inside and that fact makes the difference on the step of averaging.

3) the lower is time scale (after QPO period), the higher is the number of intervals for R/S calculation, and lower is R/S value for intervals with QPO peaks and the mean log(R/S) is proportional to log(N/2) where N is number of points on each time interval.

Thus, to find the presence of QPO in time series we don't need to perform Fourier analysis. Instead we can calculate several points in log(R/S) – log(N/2) diagram on time scale lower then period of QPO and get the value of Hust exponent using least-squares method.

QPO can naturally appear in the framework of the short noise model under the assumption that individual shots are not completely independent [5]. Under this assumption we prepared a modeled series of light curves with Poisson distributed peaks, but the form of peaks were a hight peak with exponential decay on several next time bins. The overlapping of peaks gives the power spectrum similar to the spectrum of black holes ([5]), but the fractal dimension of obtained time series was strongly connected with "dead time" interval between peaks. The founded connection requires additional studies.

The main advantage in using fractal techniques in order to search QPO in array of time series is requirements of minimal observation length: based on our observations, fractal technique can get the periodic structure in data on the time scale of several QPO periods, while for statistically good power spectrum one require several time series with the same length. This efficiency allow to change the observational program in the process of observation, if the object in field of view shows the symptoms of changed fractal dimension and consequently has the QPO in power spectrum.

4. Conclusions

The following can be concluded from our analysis:
1) the R/S analysis can be applied for studding QPO in Black Holes;
2) the average case performance of R/S algorithm is O(n), while performance of Fourier transform is O(n log(n)), that lead to the decreasing of required time especially for big data arrays.
3) every point, calculated by R/S algorithm, is calculated independently that fact give potential for parallel computing;
4) the proposed approach can be used to correct observational program in the process of observation to obtain more data from interesting objects with QPO;
5) the same approach can be used in other fields of time series analysis to find structured parts in big data arrays especially in cases when the data consist of colored noise.

References

[1] Prosvetov A V and Grebenev S A 2014 Fractal Analysis of Light Curves for the Microquasar GX 339-4 *Astronomy Letters* **40 (11)** p 691–703
[2] Hurst H E 1951 *Trans. Am. Soc. Eng.* **116** paper 2447 p 770–808
[3] Stoev S, Taqqu M S, Park C, Michailidis G and Marron J S 2006 *Computational statistics & data analysis* **50 (9)** p 2447–2471
[4] Leadbetter M R, Lindgren G and Rootzen H 1983 *Extremes and Related Proporties of Random Sequence and Processes* (Springer, New York)
[5] Vihlinin A, Churazov E and Gilfanov M 1994 Quasi-periodic oscillations in the shot noise model of X-ray sources *Astron. Astrophys.* **287** p 73–79

IOP Conf. Series: Journal of Physics: Conf. Series **913** (2017) 012009 doi:10.1088/1742-6596/913/1/012009

Block cipher based on modular arithmetic and methods of information compression

S Krendelev, N Zbitnev, D Shishlyannikov and D Gridin

Novosibirsk State University, JetBrains research

e-mail: s.f.krendelev@gmail.com, nikita.zbitnev.a@gmail.com, dmitry.shishlyannikov.a@gmail.com, dmitry.gridin.v@gmail.com

Abstract. The article focuses on the description of a new block cipher. Due to the heightened interest in BigData the described cipher is used to encrypt big volumes of data in cloud storage services. The main advantages of the given cipher are the ease of implementation and the possibility of probabilistic encryption. This means that the text encryption will be different when the key is the same and the data is the same. So, the strength of the encryption is improved. Additionally, the ciphered message size can be hardly predicted.

1. Introduction

Standard block ciphers are based on bit data representation and use standard methods of encryption. They are permutation, substitution and some characteristics of Galois fields over the field of characteristic 2. The following methods:

1) Modular methods
2) Fully homomorphic encryption methods
3) Data compression methods
4) p-adic numbers methods to build a non-periodic PRNG. This generator is not used in this work, because it has not yet been possible to increase the speed of the generator.

Let us turn to the description of the algorithm.

2. Algorithm description

It is necessary to encrypt a block of text $B = q_1, q_2, \dots, q_n$, which has n symbols of b-bit size each with secret key.

2.1. Key

The key is used to generate permutations and matrices and to fill in the tables of frequencies in Huffman coding. The assumed key size is:

$$\mathrm{l} = 2 * \mathrm{b} * \mathrm{n} \tag{1}$$

The key itself is presented as a sequence of integers: $\beta_1, \beta_2, \dots, \beta_k, \beta_{k+1}, \dots, \beta_l$, where $k = \frac{l}{2}$.

To extract all the necessary data from the key we would use an additive generator [1] (hereinafter the Generator) represented as a recurrence relation of the following type:

$$X_{k+i} = (\beta_1 * X_{k+i-1} + \beta_2 * X_{k+i-2} + \dots + \beta_k * X_i) \bmod 251, \tag{2}$$

where $X_k, X_{k-1}, \dots, X_1 = \beta_l, \beta_{l-1}, \dots, \beta_{k+1}$.

2.2. *Preparation Phase*

2.2.1. Module Choice. Let us choose a set of prime integers $M_1, M_2, \dots, M_m$ such as:

$$M = M_1 \cdot M_2 \cdot \dots \cdot M_m \in [2^b, 2^{b+1}] \quad (3)$$

It is necessary for having unique decomposition any symbol of *b*-bit size in remainders by these numbers [2].

2.2.2. Generation of Permutations. To provide nonlinear encryption it is necessary to obtain permutations for arithmetic operations for each module from the secret key [3]. To the module mod a permutation is a set of prime integers of the type:

$$\langle a_0, a_1, a_2, \dots, a_{mod-1} \rangle, \text{ where } a_i \in [1, mod - 1] \text{ and } a_i \neq a_j, i \neq j \quad (4)$$

The set is obtained by induction as:

- $a_0 = 0$;
- $a_{k+1} = a_i$, where i – is a number obtained by Generator and taken remainder of $k + 1$;
- $a_i = k + 1$.

A permutation changes addition and multiplication tables for the given module as follows. Let us take addition as an example. In a standard addition table

$$x + y = r \quad (5)$$

And in the table we managed to generate

$$x + y = i, \text{where } a_i = r \quad (6)$$

Multiplying is done analogically.

2.2.3. Generation of Matrices. For each module invertible matrices $A_1, A_2, \dots, A_m$ of $n \times n$ size are generated. To do this we use the following recurrence relation (within the framework of generated addition and multiplication tables):

$$z_{n+k} = \alpha_1 * z_{n+k-1} + \alpha_2 * z_{n+k-2} + \dots + \alpha_n * z_k, \quad (7)$$

where $\alpha_i \in [1, mod - 1]$ integers are obtained by the Generator, and $z_n, z_{n-1}, \dots, z_1 = \begin{pmatrix} 1 \\ 0 \\ \vdots \\ 0 \end{pmatrix}, \begin{pmatrix} 0 \\ 1 \\ \vdots \\ 0 \end{pmatrix}, \dots, \begin{pmatrix} 0 \\ 0 \\ \vdots \\ 1 \end{pmatrix}$.

So, let us generate using this algorithm $k = 1 + \sum_{i=1}^{n} \alpha_i$ vectors. The combination of the latter *n* vectors is a desired matrix. The obtained matrix is invertible as a matrix obtained from basic data is invertible (an identity matrix), while the following matrices are obtained by linear combination.

2.2.4. Generation of Huffman Table. At a stage of ciphering Huffman tables are used to improve the strength of the algorithm cryptography. This choice may be justified by the fact that the combination of Huffman coding [4] and insertion of fake symbols which are described below leads to the changes in the size and content of the encrypted message at every other ciphering though the key is not changed [5]. This helps to resist entropic methods of cracking and known-plaintext attacks.

Building a Huffman table and using it is as follows:

Let us build a frequency table for every symbol. As in our case all ciphered symbols will not exceed the maximum absolute value, it will be enough to build a frequency table for digits from 0 to the maximum module value. It is impossible to count real frequency in the text, because we work with only block of this text so, the frequency of every symbol will be generate by the Generator.

We generate a Huffman tree based on this table:
1) The symbols of the input alphabet make up a list of free nodes. Every node contains the weight represented in the table.
2) Two free nodes of the least weight in the tree are chosen.
3) The weight of the generated parent node is set to the sum of the weight of the children.
4) The parent is added to the queue and two nodes as children are removed from it.
5) Bit '1' represents the left branch growing out from the parent and bit '0' – the right one. Bit marking of the branches growing out from the root does not depend on the weights of 'descendants'.
6) This process is repeated until only one node remains, which is the root of the Huffman tree.

To find out the code for every symbol included into the message we must make our way from the root of the tree to the node representing the symbol. We collect bits while moving from branch to branch. The sequence of bits thus obtained is the code of the given symbol.

2.2.5. Fake Symbols. As mentioned above, the algorithm has a stronger cryptography if encrypted data look different at every other ciphering with the same key. We use fake symbols which change the look of the encrypted message for this purpose.

It is evident that if the size of the symbol is b-bit, the encrypted symbols cannot have a larger size. Also, according to the Chinese remainder theory [2], from the remainders of the division of an integer we can collect a number which is smaller than the multiplication of all modules M. Then, if a number is added into a block would satisfy following condition

$$\lambda: 2^{\mathrm{b}} < \lambda < M \tag{8}$$

it will be clear that the number is not a symbol of the source file when the block is deciphered and collected.

To implement it, we will produce a new Generator. The starting data for the new Generator are taken from system time as follows: time and date are represented at the moment of ciphering as a format string – *Www Mmm dd hh:mm:ss yyyy*, where *Www* is an abbreviated notation for a day of the week, *Mmm* stands for a month, *dd* is a date, *hh:mm:ss* is hours, minutes and seconds respectively. Obtaining a numeric representation for each symbol of the string, we have a set of 24 integers. Now these integers are used to initialize the Generator with the above-described algorithm.

Before block reading, we will obtain a number λ, from the new Generator and find the number $\lambda_1 = \lambda \, mod \, n$. If $\lambda_1 \neq 0$, then λ_1 - is a position into which a fake symbol will be put into the red block. After that read $n-1$ symbol from the text and filled the block with the fake symbol λ in position λ_1.

If $\lambda_1 = 0$, then the fake symbol is not added, n symbols are read and the block is filled in the usual way.

3. Ciphering

We will choose modules: $M_1, \dots, M_m$. Secret key is used to generate permutations, matrices $A_1, A_2, \dots, A_m$ for each module and a Huffman table. Read the block B of size $n * b$ and ciphered as follows:

1) The block $B = q_1, q_2, \dots, q_n$ is represented in the form of a vector

$$\mathrm{B} = \begin{pmatrix} \mathrm{q}_1 \\ \mathrm{q}_2 \\ \vdots \\ \mathrm{q}_\mathrm{n} \end{pmatrix} \tag{9}$$

The coordinates of the vector is remainders of $M_1, \dots, M_m$. We would have *m* vectors:

IOP Conf. Series: Journal of Physics: Conf. Series **913** (2017) 012009 doi:10.1088/1742-6596/913/1/012009

$$\mathrm{v_1} = \begin{pmatrix} \mathrm{v_{1_1}} \\ \mathrm{v_{1_2}} \\ \vdots \\ \mathrm{v_{1_n}} \end{pmatrix}, \mathrm{v_2} = \begin{pmatrix} \mathrm{v_{2_1}} \\ \mathrm{v_{2_2}} \\ \vdots \\ \mathrm{v_{2_n}} \end{pmatrix}, \dots, \mathrm{v_m} = \begin{pmatrix} \mathrm{v_{m_1}} \\ \mathrm{v_{m_2}} \\ \vdots \\ \mathrm{v_{m_n}} \end{pmatrix}, \text{where } \mathrm{v_{i_j}} = \mathrm{q_j} \bmod \mathrm{M_i} \tag{10}$$

2) Matrices are multiplied by the relevant vectors (in every module), all operation we do by tables of multiplication and addition generated previously.

$$\mathrm{w_i} = \mathrm{A_i} \cdot \mathrm{v_i}, \text{where } \mathrm{i} = 1..\mathrm{m} \tag{11}$$

3) The resulting vectors are recorded a special order in accordance with the Huffman table. The order is defined by the sequence of numbers $P = \{p_1, p_2, \dots, p_m\}$, where $p_i \in [1, m]$ – position of this module, $p_i \neq p_j \ if \ i \neq j$. So, the vectors are recorded as a sequence: $\mathrm{v_{p_1}}, \mathrm{v_{p_2}}, \dots, \mathrm{v_{p_m}}$.

When moving from block to block, matrices, Huffman tables and the order of record of the encrypted vectors are changed. Matrices are changed through string/column exchange operations or transpositions to save their invertibility. To do this we get integers c and d from the Generator. Then the decision is made which operation to perform by calculating the remainders of the division c by 3:

- If $c \ mod \ 3 = 0$, then the we use permutation of columns $c \ mod \ n$ and $d \ mod \ n$ in matrices $A_1, A_2, \dots, A_m$.
- If $c \ mod \ 3 = 1$, then the we use permutation of strings $c \ mod \ n$ and $d \ mod \ n$ in matrices $A_1, A_2, \dots, A_m$.
- If $c \ mod \ 3 = 2$, then the we use transposition of matrices $A_1, A_2, \dots, A_m$.

The order of record of the encrypted vectors is changed through the right circular shift of P to the right to number c obtained from the Generator.

The Huffman table is changed similarly. The symbols in the table remain their positions and their relevant codes are circularly shifted by the number obtained from the Generator.

4. Deciphering

We will choose modules: $M_1, \dots, M_m$. Secret key is used to generate permutations, matrices $A_1, A_2, \dots, A_m$ for each module and a Huffman table. Inverse matrices $A_1^{-1}, \dots, A_m^{-1}$, all operation we do by tables of multiplication and addition generated previously.

1. Read m encrypted vectors $\boldsymbol{v'_1}, \dots, \boldsymbol{v'_m}$ and every symbol is deciphered in accordance with a Huffman table.

2. An invertible matrix is multiplied by the relevant vector, all operation we do by tables of multiplication and addition generated previously:

$A_i^{-1} \times \boldsymbol{v'_i}, i = 1, \dots, m,$

Result of this operation is deciphered vectors $\boldsymbol{v_1}, \dots, \boldsymbol{v_m}$.

3. According to the Chinese remainder theory the obtained vectors are collected into the block B. This is a deciphered block. If it has a symbol the numerical value of which is more than 2^b, the symbol is assumed as fake and is thrown out of the block.

When moving from block to block, the changes in the reading order and in Huffman tables are done similarly to the ciphering procedure. In case of invertible matrices string permutations are replaced by column permutations. Transposition remains unchanged.

5. Cryptanalysis

5.1. PRNG

A simplified recurrent generator is used in our algorithm. It is obviously is not a good one, but it perfectly answers our needs. This Generator has been described in a number of articles [1]. Their main finding to which we will refer is the PRNG period. Parameters:

1) Module – 251, a prime number.

2) Recurrence relation length – 8. So, the estimated period is $251^8 \approx 10^{18}$.
3) At the pre-ciphering phase the Generator is called $8 + \sum_{i=1}^{m} M_i + \max_{i \in [1,m]} M_i$ times.
4) At the ciphering phase the Generator is called four more times.

Estimation: The Generator is called about 10^7 times for a file of 1 GB in size, and the Generator period is 10^{18}. It enables us to use the Generator for our algorithm.

5.2. Huffman and Influence of Fake Symbols

Fake symbols added to the block before ciphering do not influence occurrence of irrepresentable integers in encrypted vectors (an integer more than 2^{16}, when the size of symbol is 16 bits). The probability of their occurrence is always equal and depends only on the choice of a module set:

$$\frac{\prod_{i=1}^{m} M_i - 2^b}{\prod_{i=1}^{m} M_i} \approx 45\% \qquad (12)$$

for modules of: 5, 7, 11, 17, 19 and size of symbols 16 bits.

5.3. Avalanche effect

Any bit in any block is changed, the whole block changes by 40 percent on average (based on available statistics). This is very close to the definition of Avalanche effect [5]. When a bit is changed, its relevant coordinate in vectors in modules changes. As all coordinates take part in multiplication of matrix by vector, any change of a coordinate leads to a complete change of a resulting vector.

5.4. Modificated Brute Force

The classical Hill cipher is vulnerable to a known-plaintext attack. To force the attacker to do more work to break our algorithm we use the permutations of addition and multiplication tables. To restore i-th matrix after the permutations we need $(M_i - 1) * n$ texts, where $M_i -$ i-th module. Then a multiplication table must be built up and the permutations must be restored. This gives rise to two cases:

1) If all the elements occur in the matrix, the table is restored by using

$$(M_i^2 + 1) * 2^{-1} - 2 * (M_i - 1) \qquad (13)$$

texts.

2) If k elements are absent, an extra text $(k - 1) * M_i$ is needed.

Using a Huffman table helps hide vectors. First, it is necessary to restore the Huffman table or the ciphering tree to recognize vectors by the encrypted text. So, if the tree-depth is l there are $C_{2^l}^{M_{max}}$ ways to allocate M_{max} elements, where M_{max} – is a maximum module. After the table/tree is restored symbols and ciphers must be compared. The total number of variants is $M_{max}!$.

The resulting formula to crack our algorithm is:

$$C_{2^l}^{M_{max}} * M_{max}! * \prod_{i=1}^{m} \left((M_i - 1) * n * \left(\frac{M_i^2+1}{2} - 2 * (M_i - 1) \right) \right) \qquad (14)$$

The strength of the encryption for a set of modules (5, 7, 11, 17, 19) and a 6-bit Huffman code as the maximum size may be evaluated as 2^{181}.

6. Performance

A series of tests helped us find the optimum sizes of symbols and blocks. They are 16 bits for a symbol and 4 symbols for a block. Let us examine the performance of our algorithm:

6.1. Tests
Modules: 5, 7, 11, 17, 19.

Table 1. Results of tests

File size (MB)	Number of blocks	Number of fake symbols	Encrypt time (s)	Decrypt time (s)	Encrypt file size (MB)
1	131 072	138 210	0.21	0.28	1.7
32	4 194 304	4 233 113	7.02	8.72	54.05
1024	134 217 728	141 451 136	217	268	1 744

The average speed of ciphering is 5 566 634 blocks per second. It takes 790 operations to cipher one block. Thus, the performance is $4{,}4 * 10^9$ operations per second. Taking into account program parallelism, the performance can be increased m times, where m is a number of modules.

7. Conclusion
In this paper, we create new fully probabilistic encryption algorithm for working with Big Data. Probabilistic encryption guarantees high cryptographic strength.
Further research will be related to the development of random number generators not having a period and using MTF (Move-To-Front) technology from the theory of data compression [4], to generate permutations.

References
[1] Dessai V, Ravindra P and Dandina R 2012 Using Layer Recurrent Neural Network to Generate Pseudo Random Number *Sequences IJCSI International Journal of Computer Science Issues* **9** p 324–334
[2] Vinogradov I M 1954 *Elements of Number Theory* (5th ed Kravetz S, Dover)
[3] Farmanbar M and Chefranov A 2012 Investigation of Hill CipherModifications Based on Permutation and Iteration *International Journal of Computer Science and Information Security* **10 (9)** p 18–24
[4] Nelson M 1995 *The Data Compression Book* (2nd Edition IDG Books Worldwide Inc)
[5] Schneier B 1996 *Applied Cryptography Second Edition* (John Wiley & Sons Inc)

Intelligent search in Big Data

E Birialtsev[1], N Bukharaev[2] and A Gusenkov[3]

[1] R&D Director Gradient Ltd, PhD, Kazan, Russia
[2] Associate Professor, PhD, Institute of Computational Mathematics and Information Technologies, Kazan Federal University, Kazan, Russia
[3] Senior Lecturer, PhD, Institute of Computational Mathematics and Information Technologies, Kazan Federal University, Kazan, Russia

E-mail: [1] igenbir@yandex.ru, [2] boukharay@gmail.com, [3] gusenkov.a.m@gmail.com

Abstract. An approach to data integration, aimed on the ontology-based intelligent search in Big Data, is considered in the case when information objects are represented in the form of relational databases (RDB), structurally marked by their schemes. The source of information for constructing an ontology and, later on, the organization of the search are texts in natural language, treated as semi-structured data. For the RDBs, these are comments on the names of tables and their attributes. Formal definition of RDBs integration model in terms of ontologies is given. Within framework of the model universal RDB representation ontology, oil production subject domain ontology and linguistic thesaurus of subject domain language are built. Technique of automatic SQL queries generation for subject domain specialists is proposed. On the base of it, information system for TATNEFT oil-producing company RDBs was implemented. Exploitation of the system showed good relevance with majority of queries.

1. Introduction

Exponential growth of input data volume, measured today in zettabytes, led to qualitative changes in computer aided technologies of acquiring, storing, processing and analysing information known as the Big Data phenomena. Forrester Research Company [1], for instance, defines the notion as a hardware or software technology to organize, control and analyse data arrays, characterized by "four Vs" — Volume, Variety, Variability and Velocity. NoSQL [2], MapReduce [3], Hadoop [4], SAP HANA [5] are the most widely known technologies of the kind.

Availability of sufficiently effective search procedure can be stated as one of the main goals for data integration. On the other hand, volume and variety of the Big Data makes problem of integration of heterogeneous information resources non-trivial. M. Kogalovsky proposed in [6] the classification of data integration systems, based on division of physical, logical and semantic levels of integration. Integration on physical level means plain conversion of information taken from heterogeneous data sources to some common format of physical representation. Integration on logical level provides a global scheme of uniform access to such sources, describing common data representation with respect to structural and (in object-oriented models) behaviouristic properties of the data sources. Integration on semantic level supports data representation reflecting just semantic properties of information, described within framework of some subject domain ontology. The latter allows to use ontology terms as conceptual reasoning model, giving great advantage to base end user interface on high level data model. For ontology description OWL language, developed by Semantic Web Activity working group and recommended by W3C consortium [7], is usually used.

IOP Conf. Series: Journal of Physics: Conf. Series **913** (2017) 012010 doi:10.1088/1742-6596/913/1/012010

2. RDBs integration

Let's describe, in its most essential features, the following approach for RDB integration, aimed on organization of effective search in large information systems, containing some several dozens of local RDBs. Those RDBs may have different logical structure and physical representation, but are related to one subject domain. For such integration, we have to use structural information, retrieved from the databases, and more specific information on the subject domain. In other words, to solve the problem of data integration on semantic level we need to define explicitly auxiliary information resources such as physical RDBs model, logical model of subject domain and thesaurus of end user terminology, presented in the formalism of ontologies.

3. RDBs ontology

For RDBs integration we need to retrieve information about database structure. Structural problems appearing here are thoroughly analysed in [8–10]. To output meaningful information, the search engine should have information about possible table joins. Such information can be easily retrieved from database schemes, as modern RDBs store system information on keys. Note also that majority of the modern RDBs store information on possible joins to support data integrity. So, restoring all possible meaningful combinations of joining RDB tables is a fairly simple task.

There are many well developed formalisms to describe database structure, such as DDL SQL, ER-diagrams [11], UML [12] etc. Their expressive power is sufficient for solving problems of database design and data access support in the case when each database is considered separately, each one with its own system of attributes. But let's note that in the case of RDBs data integration, there arise problems, related to the peculiarities of naming of database artefacts (such as tables and fields) [13]. These peculiarities force us to treat the artefacts names not just as atomic units, but as independent semantically linked objects with their own structure.

There are several ways to describe RDB structure in terms of ontologies, which can be reduced to the following two main approaches.

The first one [14] implies

- conversion of the set of database tables to the set of ontology concepts, having the same name and slots, corresponding to the tables columns, and
- projection of joins of tables into relations between concepts in the form of migrating keys.

The set of data domains, i.e. all predefined and user defined data types, is also included into set of ontology concept. Thus, in this case table records, key values and data types form the set of instances of ontology concepts.

We prefer, as more universal one, the second approach proposed in [9, 10], which presumes higher level of the concepts abstraction. In this case RDB structure is described in terms of universal concepts such as

- TABLE, COLUMN, KEY, DOMAIN, corresponding to the main RDB objects, and
- General relations between the objects such as
  - TABLE contains COLUMN,
  - TABLE has primary KEY,
  - TABLE has foreign KEY,
  - COLUMN contains KEY and
  - COLUMN has type DOMAIN.

Two following interpretation functions are also introduced [15]:

φ1: If TABLE1 has primary KEY1 and TABLE2 has foreign KEY1, then there exists TABLE3, containing all columns of TABLE1 and TABLE2.

φ2: If TABLE1 contains COLUMN1, then there exists TABLE2, containing all columns from TABLE1 except COLUMN1.

The first function corresponds to the table join operation, and the second one corresponds to the relation projection operation. The second function is used to select desired columns from table joins. The data integration problem then is reduced to the problem of finding different ways to extract the specified attributes (columns) from the tables of the RDBs — i.e. finding such sequence of φ1 and φ2 functions applications, that gives in the end desired set of columns from the given ontology. This problem belongs to the well-known class of problems on a walking along the oriented graph, in which vertices represent instances of TABLE concept and edges reflect presence of the one-to-many relation. Methods of solving this class of problems are well known [16], although increase of computation complexity depending the problem dimension should be taken onto account.

4. Subject domain ontology

Creating ontology is a time-consuming process, which usually is carried out by a team of specialists, highly qualified in computer linguistics and in specific subject domain. Let's describe here technique proposed in [9, 10, 17, 18] for oil-producing ontology creation.

Oil industry standard EPICENTRE logical model, developed by Petrotechnical Open Software Corporation (POSC) [19], was chosen as an initial ontology prototype. The model is represented in the form of the set of ER-diagrams considered together with the set of text files on EXPRESS object-oriented language. It defines more than 1000 real technical and business objects (or entities, in the POSC terminology), related to the field of oil exploration and production.

As usual in the object-oriented approach, the objects here are characterized by their attributes. The most important objects attributes define relationships between entities. Object-oriented concept of inheritance is also an essential part of the EPICENTRE architecture, providing effective way of organizing extremely large set of entities into a compact logically linked structure.

In our approach OWL-DL dialect of the OWL language [7], developed by Semantic Web Activity working group and recommended by W3C consortium, has been chosen as a unified ontologies representation language. The core of the approach is the scheme of conversion of the EPICENTRE model to its OWL-DL description. To implement the scheme, LR(1) formal grammar of the EPICENTRE model was defined. The Russification of the description of the EPICENTRE entities and attributes, as well as all corresponding classes and properties on OWL, was also performed.

5. Linguistic thesaurus

Creation of the linguistic ontology of natural and technical objects follow the main principles of the WordNet thesauri construction [20]. The dictionary of the subject domain is constructed by combining word phrases from descriptions of entities and attributes of the EPICENTRE model with word phrases taken from descriptions of tables and domains attributes of the TATNEFT oil-producing corporation RDBs. The lexical and semantic characteristics of the databases, used in constructing the dictionary, are described in detail in [9, 21].

For each word phrase an input set of synonyms or synset is defined. On the set of lexical-semantic variants of word phrases and synsets, the following linguistic relationships are defined: hyponymy, part-whole, incompatibility, antonymy, converseness, homonymy. The synsets of the thesaurus constructed in this way inherit the features specific to real descriptions of the RDB attributes, such as the presence of short phrases, general and technical abbreviations, spelling errors. Note however that, unlike the analysis of texts in natural language, it is possible to refine the recognition of input words by viewing the contents of the tables attribute descriptions and comparing them with the subject domain ontology.

6. Natural language search and SQL queries generation

In this section, we describe in brief a way of automatic generation of SQL queries, which requires from the end users to be knowledgeable just in the subject domain. This technique uses subject domain ontology, instance of RDB ontology and thesaurus of user terminology described earlier. At first, we should stress here that exactly the automation of the subject domain ontology construction, the linguistic thesaurus and universality of the RDB ontology makes described approach usable by Big Data collection, storage and processing technologies [1, 2]. Note also that automation of the mentioned objects construction allows us to create information systems more resistant to modifications in process of usage. In case when the logical model of the subject domain is changed, the subject domain ontology can be reconstructed and the linguistic thesaurus can be replenished quite easily.

From existing implementations, let's point to the InBase system, developed by the school of A. S. Narinyani [22] as the closest to the proposed approach. A distinctive feature of the InBase is usage of advancing semantic analysis for parsing and understanding queries. The analysis here is based on the object model of the subject domain, linked by the designer to the RDB model. This approach has a number of significant advantages in comparison with the syntactic templates usage. In particular, it eliminates the need to specify in the query patterns all word forms and all possible orders of words.

At the same time, it should be recognized that the InBase system has not found wide application yet. As it seems to us, the approach developed in InBase system has a number of essential problems. The most significant of them is the interpretation of RDB columns as attributes of classes of the object model. Real databases quite often implicitly refer to the complex notions, that can be hardly adequately mapped to the object model — since they rather mean a subquery rather than a classic attribute. This issue has been considered in detail in [13, 21], so let's here confine ourselves just to a single counterexample. Say, in one of the explored real industrial RDBs there were columns with comments like "The well production rate before repair" and "The well production rate after repair", which cannot be described as class attributes.

The second problem is necessity to bind the columns to the object model attributes manually. In addition to the great complexity of this process during the initial implementation, it also creates difficulties during system maintenance. In fairly often cases of changing RDB structure, labor intensive adjustments of the object model and its projection to the tables columns are required.

In corporate information systems, RDBs can contain hundreds of tables and thousands of columns. The investigation of the question which specific database, table and columns contains the required information can take several days, whereas the formulation of the query in natural language by the users knowledgeable in the subject domain will take several minutes. Note also that in such cases no specialist uses the natural language in all its richness, but is restricted to some simplified professional dialect.

In the proposed approach, information processing is based on the following principles:

- Dialogue between the system and the user is carried out in a table form;
- Semantic approach to columns search is used;
- Search of possible table joins by database keys is automated;
- Visual procedures for defining table selection operations are used.

The end user defines queries in the form of a table in natural language, using the terms of the thesaurus (subject domain language). See table 1 for an example of such query. Note that the query, formulated as a sentence in natural language, would be very cumbersome not only for computer analysis, but also for human understanding.

Table 1. Query example

Column Name	Condition
Well No.	10
Date of commissioning	> 1.06.2001
Date of overhaul	
Expected oil production rate	>0
Actual oil production rate	

IOP Conf. Series: Journal of Physics: Conf. Series **913** (2017) 012010 doi:10.1088/1742-6596/913/1/012010

Recall that semantic information is presented in the system as a semantic network (ontology), containing a set of possible links between the concepts. Semantic parsing of the RDB columns comments is an almost completely automatic operation, which allows to quickly attach a large corporate database to the subject domain ontology. Semantic analysis of the user's query is reduced to identifying the subgraphs of the network, which connect the concepts, corresponding to words phrases, used in the columns names.

Algorithm of semantic analysis of the user's query is revealing the subgraphs in the taxonomic tree of the subject domain ontology, connecting the concepts named by word phrases from the linguistic thesaurus, corresponding to the query columns (see figure 1).

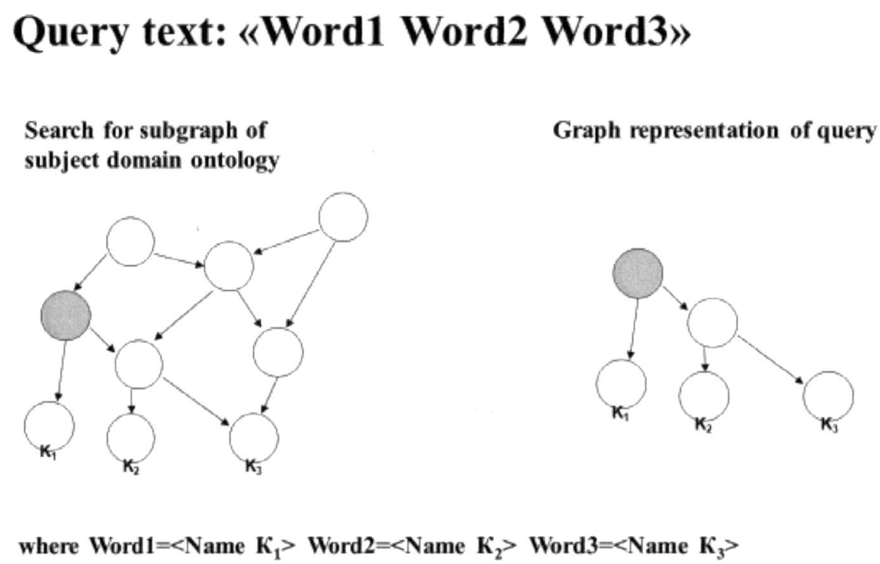

Figure 1. Understanding of query in ontology context

7. Implementation of the system for Tatneft

In this section, we briefly outline the functionality of the web-based search system implemented in accordance with the approach described above for Tatneft oil-producing company. The system is described in more detail in [9, 10, 23, 24].

For user administration, a standard password entry scheme for individual users is used here. During the login process, each of the users is assigned one of three roles — User, Technologist or Administrator.

7.1. End user subsystem

Using the system's tools, the user constructs a query in the form of a table (see example in table 1). The names of the columns are formulated in a dialogue with the system in an arbitrary form with the use of phrases from the linguistic thesaurus. The number of columns is unlimited.

After setting the query, the user sends it to the generation. The system performs a semantic analysis of the query, builds the SQL expression and evaluates the query execution time to inform the user. The built-in SQL query is presented to the user, who can edit or run it for execution. The user interface is simplified as much as possible to allow any user with an understanding of the subject domain to work with the system. Knowledge of SQL language and database structures are not required for the average user.

For most queries, the system has shown good relevance of the results.

7.2. Technologist subsystem

This subsystem for flexible configuration of the system is of particular interest for evaluating the advantages of the approach. The main goal of a technologist is to mark up the columns of the database in a way that ensure the maximum relevance of the end user queries. To accomplish this task, he can use the following built-in tools:

- Subject domain language (thesaurus) editor;
- Subject domain ontology editor;
- Database ontology editor;
- Search algorithm configuration tools;
- Tools for analysis of completeness and adequacy of ontologies.

The system provides the built-in lexical editor to allow technologists to introduce new terms into the thesaurus. Note here that the presence in the user's query of words absent in the thesaurus does not cause fatal consequences — an unknown word is marked, but it is ignored in the future analysis.

The ontology editor of the domain allows technologist to edit a specific concept or define a new one as well as to specify, delete or change the links of this concept with other concepts. Each link can be customized individually by changing the semantic distance between the concepts. This tool is also used to fine-tune the search engine at the stage of marking the names of user-defined columns and their matching with comments.

Technologists have an access to the list of all RDB tables, together with service information on the current state of their markup by concepts of the subject domain ontology. The subsystem allows them to specify the settings for the databases with which the system can work simultaneously. Thus, the subsystem allows to fulfil all necessary actions for tuning the system when determining the needed RDBs or changing the subject domain terminology are necessary.

7.3. Administration subsystem

The subsystem provides implementation of standard administrative functions such as maintaining the user registry, determining the access rights to work with system objects, logging user actions, uploading and uploading the ontology. It also supports direct manipulation with the service tables by means of SQL.

8. Conclusions and perspectives

Practice of implementation and numerous experiments show fruitfulness of the described approach to the data integration and search in structurally marked texts, which are seen here as multiply connected graphs, implicitly reflecting the semantics of the subject domain.

Defining a query, an expert user presumes some meaningful links between key words used in the definition. Thus, formulation of the query as well as structure of stored information implicitly reflect semantics of the subject domain. This combination helps the search engine to find complete and sufficiently relevant set of answers. That explains effectiveness of quite simple search algorithm implemented in described approach.

At the same time, the considered search mechanism has several obvious directions for further development. Say, important results could give analyses of real mechanisms of data normalization. The known normal forms, including the generalized normal form proposed by R. Fagin in 1981, do not change the set of the database attributes, but just redistribute them between tables. However, in the real databases, quite often a key with a small number of values is transformed to a group of attribute names or tables. Thus, the set of attribute values and the set of attribute names can be transformed into each other. De facto, this practice is reflected in some constructions of storage mechanisms of industrial databases — for example, in segmentation of large tables by keys, supported in Oracle, PostgreSQL and a number of other systems. To some extent, this practice has been also legitimized in repositories with the capability of integration of data from various sources, where attributes can be stored in a table by introducing an additional attribute "Attribute name". Thus, the practice of transformation of the attribute name into the value of some super-type exists, but so far it hasn't received sufficient theoretical coverage. In our opinion, detailed study of this phenomenon can give a valid contribution to the theory of relational databases.

In the context of the main topic under consideration, further formalization of this restructuring phenomenon can provide a key to automating the construction of queries and data replication between transaction systems with different database structures, as well as with repositories, multidimensional OLAP systems and other storage formats, including those based on XML. Creation of search engines, based on such new theoretical principles, is especially important for Big Data, for which heterogeneity of information representation is inherent.

A In the context of the main topic under consideration, further development of such restructuring mechanisms can provide a key to automating the construction of queries and data replication between different databases, as well as between repositories, multidimensional OLAP systems and other storage formats, including those based on XML. Creation of search engines, based on such new theoretical principles, is especially important for Big Data, for which heterogeneity of information representation is inherent.

In the context of the main topic under consideration, further development of such restructuring mechanisms can provide a key to automating the construction of queries and data replication between different databases, as well as between repositories, multidimensional OLAP systems and other storage formats, including those based on XML. Creation of search engines, based on such new theoretical principles, is especially important for Big Data, for which heterogeneity of information representation is inherent.

An obvious possible obstacle to the described approach is put by its combinatorial complexity. However, continuously increasing data processing power can to great extend overcome this barrier.

Acknowledgments
This work was funded by the subsidy allocated to Kazan Federal University for the state assignment in the sphere of scientific activities, grant agreement no. 1.2368.2017.

References

[1] Hopkins B and Evelson B 2012 *Expand your digital horizon with Big Data* Avaliable from: http://www.asterdata.com/newsletter-images/30-04-2012/resources/forrester_ex-pand_your_digital_horiz.pdf
[2] *NOSQL Databases* Avaliable from: http://nosql-database.org/
[3] *Apache Hadoop, MapReduce* Avaliable from: https://hadoop.apache.org/docs/r1.2.1/map-red_tutorial.html
[4] *Apache Hadoop* Avaliable from: http://hadoop.apache.org/
[5] *SAP HANA* Avaliable from: https://www.sap.com/products/hana.html
[6] Kogalovskij M 2010 *Methods of data integration in information systems* (Moscow: Russian Academy of Science Institute of Market Problems) Avaliable from: http://www.ipr-ras.ru/articles/kogalov10-05.pdf
[7] *W3C, OWL Web Ontology Language* Avaliable from: https://www.w3.org/TR/2004/REC-owl-features-20040210/
[8] Birialtsev E, Gusenkov A and Kosinov Y 2007 Representation of relational database structure in the ontology formalism *Proc. of Kazan School on Computer and Cognitive Linguistics TEL-2006* (Kazan: Otechestvo Publishing House) p 32–37
[9] Birialtsev E and Gusenkov A 2007 Integration of databases based on ontologies *Scholarship Notes of the Kazan University. Physics and mathematics series* **149** p 13–25
[10] Gusenkov A, Birialtsev E and Zhibrik O 2015 *Intellectual search in structured information arrays* (LAP LAMBERT Academic Publishing)
[11] Garcia-Molina H, Ullman J and Widom J 2003 *Database Systems: The Complete Book* (Williams Publishing House)

[12] Booch G, Rumbaugh J and Jacobson I 2000 *The Complete UML Training Course* (Moscow: DMK Press)
[13] Birialtsev E, Gusenkov A and Galimov M 2005 Features of the lexico-semantic structure of the relational databases artifacts naming *Proc. of Kazan School on Computer and Cognitive Linguistics TEL-2005* p 4–12
[14] Zhuchkov A, Arnautov S and Tverdohlebov N 2003 New technologies for conceptual networks created within framework of the project "New generation vaccines and diagnostic systems of the future" *Electronic Libraries* **6** Avaliable from: http://www.elbib.ru/index.phtml?page=elbib/rus/journal/2003/part6/ZATGS
[15] Gavrilova T and Horoshevskij V 2001 *Knowledge base of intelligent systems* (St. Petersburg: Piter Publishing House)
[16] Anderson J 2003 *Discrete Mathematics with Combinatorics* (Moscow: Williams Publishing House)
[17] Birialtsev E, Gusenkov A and Khairullina A 2006 Presentation of the POSC Epicenter data model in the OWL ontology language *Proc. of Kazan School on Computer and Cognitive Linguistics TEL-2006* p 38–49
[18] Birialtsev E and Gusenkov A 2007 The construction of subject domain ontology based on the logical database model *Proc. of Knowledge-Ontology-Theory (ZONT-07) Conf.* (Novosibirsk: Sobolev Mathematical Institute Press 1) p 176–183
[19] *Energistics, Epicentre v3.0* Avaliable from: http://www.energistics.org/energistics-standards-directory/epicentre-archive
[20] Fellbaum C (ed.) 1998 *WordNet: An Electronic Lexical Database* (Cambridge: MIT Press)
[21] Birialtsev E and Gusenkov A 2007 Relational database ontologies. Linguistic aspect *Proc. Of Dialog'2007 Conf.* P 50–53
[22] Narinyani A S 2001 Centaur named Theon: Thesaurus + Ontology *Proc. of the international workshop on computational linguistics and its applications DIALOG* **1** p 184–188
[23] Birialtsev E, Gusenkov A and Mironov S 2009 One approach to implementing unregulated access to relational databases *Proc. of Kazan School on Computer and Cognitive Linguistics TEL-2008* p 10–23
[24] Chaudhuri S and Dayal U 1997 An Overview of Data Warehousing and OLAP Technology *ACM SIGMOD Record* **26(1)** p 65–74

IOP Conf. Series: Journal of Physics: Conf. Series **913** (2017) 012011 doi:10.1088/1742-6596/913/1/012011

Efficiency Analysis of the access method with the cascading Bloom filter to the data warehouse on the parallel computing platform

Yu A Grigoriev, V A Proletarskaya, E Yu Ermakov and O Yu Ermakov

Bauman Moscow State University

e-mail: grigoriev@bmstu.ru, JK.Ermakov@gmail.com, vilka2000@mail.ru, iHelos.Ermakov@gmail.com

Abstract. A new method was developed with a cascading Bloom filter (CBF) for executing SQL queries in the Apache Spark parallel computing environment. It includes the representation of the original query in the form of several subqueries, the development of a connection graph and the transformation of subqueries, the definition of connections where it is necessary to use Bloom filters, the representation of the graph in terms of Spark. On the example of the query Q3 of the TPC-H test, full-scale experiments were carried out, which confirmed the effectiveness of the developed method.

1. Introduction

Relational databases (RDBs) still retain their leading position in many areas of data processing. But with the advent of technologies for processing large amounts of data (Big Data), the RDB's shortcomings began to manifest itself: limited scalability, the impossibility of storing aggregates in explicit form, etc. A new class of systems, known as NoSQL [1], appeared to overcome the limitations of RDBs.

NoSQL databases have high horizontal scalability (thousands of nodes) and eliminate some limitations of the relational model (they allow storing aggregates explicitly, there is no single database schema, etc.), but they do not support the conduct of transactions, which is important for financial systems. To overcome this lack of NoSQL, the NewSQL databases [2] appeared. They support a relational model, including transaction management, and at the same time have high horizontal scalability. But NewSQL inherits the disadvantages of the relational model: you can not store aggregates explicitly, you need to develop a single database schema.

Following NoSQL, platforms that support MapReduce (MR) technology [3] appeared. They provide parallel processing of queries (in particular, SQL queries) on large-scale clusters (Hadoop – 4000 nodes). Their advantage is also the large number of replicas and, consequently, the high reliability of processing huge amounts of data.

Currently, intensive work is underway to expand the capabilities of MapReduce technology. So, along with Hadoop, the new Spark software platform became very popular [4]. Like Hadoop, Spark is not just a separate project, but an entire ecosystem of various projects based on it.

There are several methods of accessing the data warehouse within the MapReduce / Spark framework, which differ in the number of tables in the query, the number of phases, the format of storing records, the distribution of data by nodes, and the use of early or late materialization of search results. A critical analysis of the existing methods was carried out in papers [5–9]. However, these methods have a limited scope of applicability, and also have a direct dependence on the amount of node memory.

The article proposes a method for accessing a data warehouse with an arbitrary schema. It is based on the cascading Bloom filter (CBF) and allows you to implement a virtually arbitrary SQL query in the Spark environment.

2. Existing methods for applying the Bloom filter to a data warehouse with a star schema

The Bloom filter [10] is a specially constructed array of bits based on hashing the key values of the dimension table. The corresponding foreign key of each record of the fact table is processed by the Bloom filter [11]. This filter gives a "misfire" with a probability of 1 / 2K, where K is the Bloom filter parameter. At a sufficiently large value of K, it is possible to achieve that the probability of false positive filtering would be very small. In this case, the number of entries participating in the connection is reduced, and in some cases it is not necessary to connect the dimension table to the fact table at all. This allows you to significantly reduce the processing time of requests.

The Bloom filter is already used in practice. Brito and others showed the effectiveness of using Bloom filters on the example of a data warehouse of the "star" type [6]. Compared to the methods implemented in MR (Fig. 1, rectangle) this method (SBFCJ – Spark Bloom-Filtered Cascade Join) wins by elapsed time and the amount of data transferred between the nodes (shuffled bytes) and additionally saved on disk (disk spill).

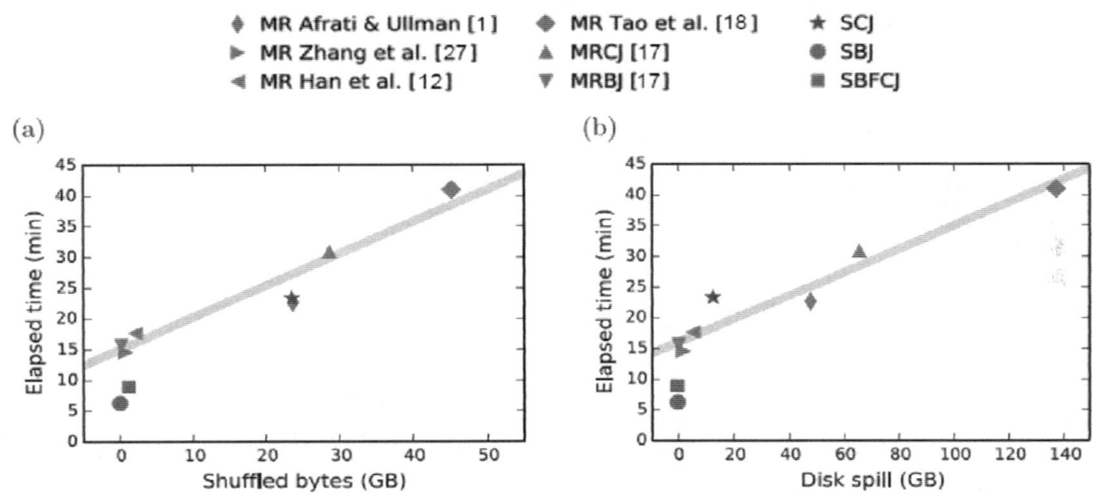

Figure 1. System response time, depending on the amount of data transferred across the network (a), and the amount of data stored on the disk due to lack of RAM (b) [6]

It is competing with the SBJ (Spark Broadcast Join) method, in which all the measurement tables are cached in RAM on each node (Fig. 1, circle). However, when using the SBJ method, strong fluctuations in the amount of used RAM are observed at sufficiently large storage sizes and small RAM volumes [6]. Figure 2 shows the dependencies of the query execution time Q4.1 from the Star Schema Benchmark (SSB) test [12] on the amount of RAM provided to one executor (Java machine).

IOP Conf. Series: Journal of Physics: Conf. Series **913** (2017) 012011 doi:10.1088/1742-6596/913/1/012011

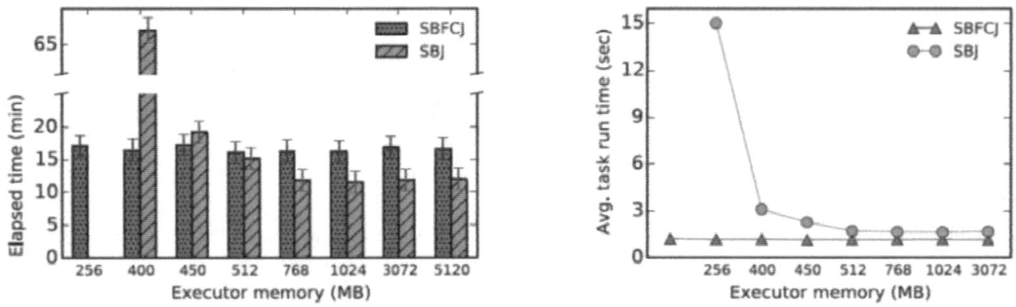

Figure 2. Comparsion the characteristics of SBJ and SBFCJ with 20 executors depending on the size of the RAM for one executor [6]: the query execution time and the average time for the task

The experiment was performed with 20 executors. The SSB test database has a star schema (4 dimensions, 1 fact table); Query Q4.1 – a connection of 4 measurements and fact tables, grouping, sorting; The request was made to fill SF = 200: 1.2 billion records in the fact table (tf), the volume of tf – 130 GB. The cluster had the following characteristics: the number of nodes – 21 (1 master and 20 slaves), Hadoop 2.6.0 and Apache Spark 1.4.1, OS GNU / Linux installation (CentOS 5.7), each node – 2GHz AMD CPUs and 8GB RAM.

When the volume of RAM for one executor is less than 400 MB, there is a sharp increase in the execution time of the query and a separate task when using the SBJ method. Thus, the SBFCJ method is more stable when the volume of the RAM changes. Therefore, the Bloom filter was selected by the authors to develop a method for parallel processing a query to a data store with an arbitrary structure.

3. Implementation of the CBF method in SPARK

The developed method includes the following steps. 1) Representation of the original query in the form of several subqueries. Each subquery performs a search in one table. 2) Development of connection graph and subquery transformation. Identify the connections where you need to use Bloom filters. 3) Representation of a graph in the language of Spark. Let us illustrate this method with concrete examples.

Example 1. Take a fairly complex Q17 query from the TCP-H test, which includes a correlated select subquery (table 1, column 1). This query determines how much the annual revenue will be lost if the orders are not sold for small quantities of certain lots.

Table 1. Query Q17

Query Q17	Subqueries
select	<u>Subquery*P1:*</u>
sum(l_extendedprice)/7.0 as avg_yearly	**select** p_partkey as pr1 **from** part
from **lineitem, part**	**where** p_brand = '[BRAND]'
where **p_partkey = l_partkey**	and p_container = '[CONTAINER]';
and p_brand = '[BRAND]'	<u>Subquery*P L1:*</u>
and p_container = '[CONTAINER]'	**select** l_partkey as pr1,
and l_quantity < (	0.2 * avg(l_quantity) as a1 **from** lineitem
select **0.2 * avg(l_quantity)**	**where** l_partkey=BF1
from **lineitem**	**group by** l_partkey;
where	<u>Subquery*P L2:*</u>
l_partkey = p_partkey	**select** l_partkey as pr1, l_quantity as q1,
);	l_extendedprice as e1 **from** lineitem
	where l_partkey=BF2;

IOP Conf. Series: Journal of Physics: Conf. Series **913** (2017) 012011 doi:10.1088/1742-6596/913/1/012011

Based on the description of the query Q17, we form the subqueries P1, L1, L2, which are presented in Table 1 (column 2). Figure 3 shows the connection graph and the subquery transformation. Here the following designations are accepted: rectangles designate the original tables; Rectangles with rounded corners – tables (RDD), which are formed by Spark after subqueries P1, L1, L2; Rectangles with dashed border – the results of joining tables of subqueries; Ovals – additional operations (filtering, aggregation). Arcs with arrows denote the initial data for the formation of tables or operations, arcs without arrows – the conditions of the connections of the corresponding tables. The circles indicate the Bloom filters (above them are the attributes for which these filters are built).

Executing the subquery P1 allows you to reduce the power of the pr1 attribute. Therefore, the use of the Bloom filters BF1 and BF2 here is appropriate.

From the description of the sequence of execution of the original query Q17, it becomes clear that Bloom compact filters are sent over the network, not the measurement tables. The join with the fact tables is performed only after Bloom filters are applied to them, i.e. with tables that are smaller in volume.

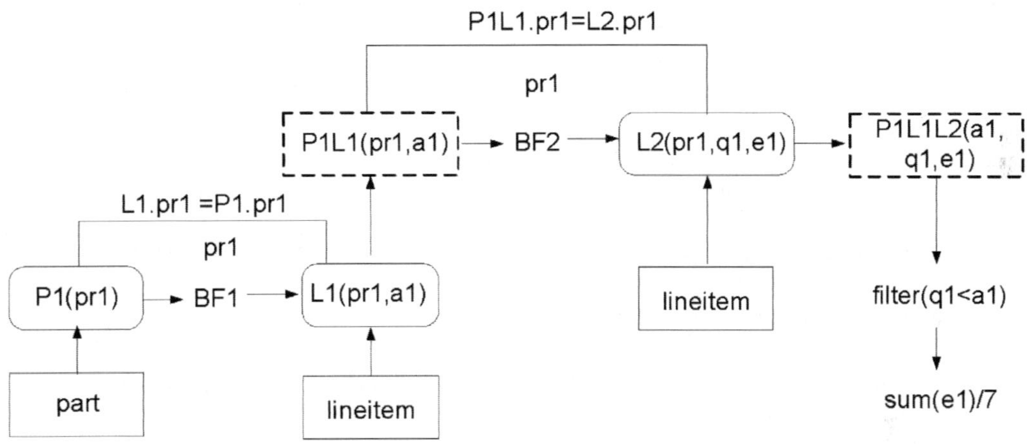

Figure 3. Connection graph and subquery conversion for Q17

Example 2. Table 2 contains a query Q3 (without a correlated subquery) from the TCP-H test and a description of its subqueries.

Table 2. Query Q3.

Query Q3	SubQuery
select **l_orderkey,**	SubQuery *C1:*
sum(l_extendedprice*(1-l_discount)) as	**select** c_custkey as kc1 **from** C_
revenue, o_orderdate, o_shippriority	**where** c_mktsegment = '[SEGMENT]';
from **customer C_, orders O_,**	SubQuery*O1:*
lineitem L_	**select** o_custkey as kc1, o_orderkey as ko1,
where **c_mktsegment = '[SEGMENT]'**	o_orderdate as dt1, o_shippriority as sh1
and c_custkey = o_custkey	**from** O_ **where** o_orderdate < date '[DATE]'
and l_orderkey = o_orderkey	and o_custkey=BF1;
and o_orderdate < date '[DATE]'	SubQuery *L1:*
and l_shipdate > date '[DATE]'	**select** l_orderkey as ko1, l_discount as d1,
group by **l_orderkey, o_orderdate,**	l_extendedprice as e1 **from** L_
o_shippriority	**where** l_shipdate > date '[DATE]'
order by **revenue desc, o_orderdate;**	and l_orderkey=BF2;

Figure 4 shows the join graph and the subquery transformation for Q3, constructed according to the rules given above.

IOP Conf. Series: Journal of Physics: Conf. Series **913** (2017) 012011 doi:10.1088/1742-6596/913/1/012011

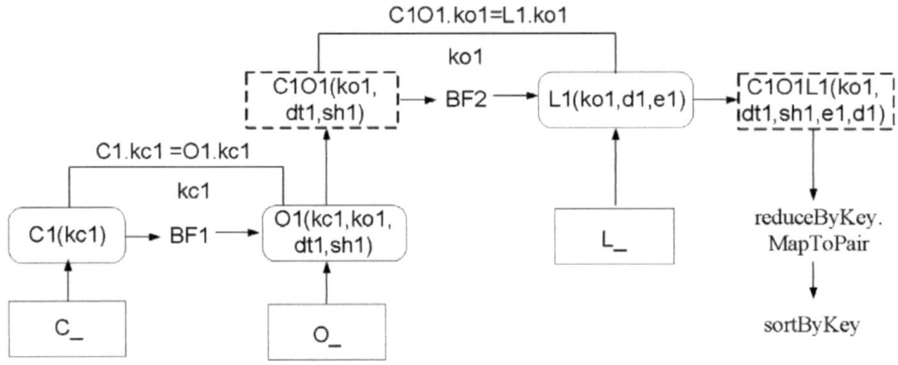

Figure 4. Connection graph and subquery conversion for Q3

In general, in each "star" can be present several tables. In this case, the original tables can be repeated in one or different snowflake nodes. For example, the table Lineitem in query Q17 is repeated in two nodes (see Table 1, column 1).

The sequence of transformations and actions in Spark forms a DAG (Directed Acyclic Graph) for the execution of the query. In the general case, the plan can be complex and have the form of an acyclic graph. It works as a pipeline, in parallel "pushing" the partitions (partitions) of RDD (fragments of tables that are stored in the nodes of the cluster) between the nodes of the graph. Figure 5 shows the plan for the "star" scheme, which can act as a measurement in another "star". The track for the promotion of the sections has the following form (in figure 5, the numbers of stages are indicated in figures).

Step 1. Reading the change tables Di, filtering the records, obtaining the projection (k_i, s_i).

Step 2. Keeping keys k_i on the local disk (before shuffle), collecting measurement keys on one Executor (for each measurement), building Bloom filters (for each measurement), broadcasting Bloom filters for all Executors where the fact table filtering (FT). This is the synchronization point for all previous processes.

Step 3. Reading the fact table, filtering records with condition F and using Bloom filters, obtaining the projection ($\{fk_i\}$, k_F, s_F),

Step 4. Serial connection of the filtered fact table with the filtered tables of measurements (Join i). Grouping and sorting, if it is provided for this "star". Go to the next "star".

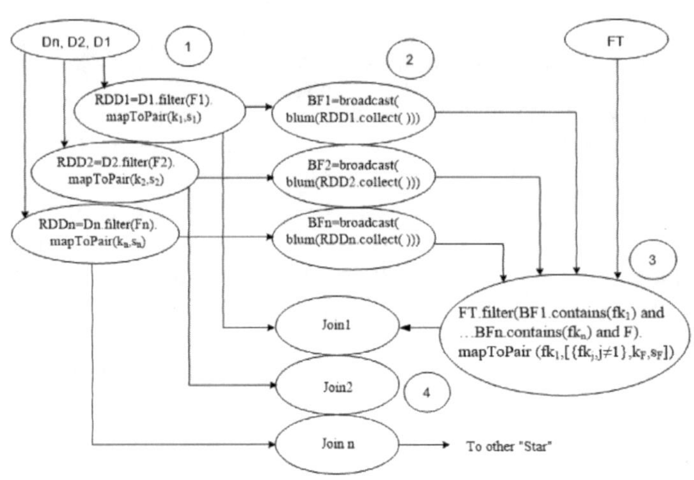

Figure 5. DAG

IOP Conf. Series: Journal of Physics: Conf. Series **913** (2017) 012011 doi:10.1088/1742-6596/913/1/012011

4. Experiment description and the result analysis

Virtual cluster was deployed to carry out an experimental comparison of the developed CBF method and the method without using the Bloom filter (NBF). The cluster consisted of 8 nodes, one of which was installed HDFS management services, Hive, Spark 2, Yarn. The main characteristics of each virtual node were as follows: one dual-core processor, 8 GB of RAM, 200 GB SSD disk, Ubuntu OS 14.16.

	NBF	Time (min.) execution, SF=500, N=40 ml.
0	Reading C_ and Blom filter bloom1 (Stage0) creating	0,4
1	Reading O_ and bloom1 (Stage1) using	1,5
2	map C_ – join with O_ (Stage2)	0,3
3	C_▶◀O_ and bloom2 (Stage3) creating	0,6
6	map C_▶◀O_ –join with L_ (Stage6)	0,1
7	Reading L_ and bloom2 (Stage7) using	7,3
8	L_▶◀ (C_▶◀O_), group by, order by (Stage8)	1,2
		SUM: 0,4+1,5+0,6+7,3+1,2=11 min

Figure 6. NBF stages

First consider the results of Q3 query execution using the NBF method (see Table 2, column 1). Figure 6 shows the stages of the NBF and the time of their implementation in Spark for SF = 500 (this corresponds to approximately 400 GB of processed data). The ▶◀ mark indicates the table join operation.

You can see that reading the source tables L_, C_, O_ is done in parallel. The query execution time was 25 minutes.

figure 7 shows the characteristics of the NBF task execution stages (SF = 500). The executor is the Java virtual machine (JVM) on which the stage tasks are performed. By default, the number of CPU cores per one instance of the executor is 1 (parameter spark.executor.cores).

SF=500 NBF Stages	Quantity			Quantile task time, sec.					Avg. time per Task	
	nodes	executors	Tasks	min	0,25	0,5	0,75	max	Garbage collection	Shuffle write
0	7	13	455	0,2	8	12	16	49	0,028	0,1
1	1	2	24	0,027	2	3	6	10	0,011	0,055
2	1	13	100	3	7	10	15	28	0,037	0,2
3	7	13	14	31	36	40	42	45	1	0,5
4	7	13	14	210	300	318	324	342	4	
9	7	13	14	144	264	276	282	288	6	0,028
10	1	1	1	1	1	1	1	1	0,026	

Figure 7. characteristics of the NBF task execution stages (SF = 500)

On fig. 7 arrows shows that the reading of the original tables (stages 0 ÷ 2) is performed in parallel on 13 executors (13 CPU cores on 7 nodes). The average number of tasks performed by one core, in this case is (455 + 24 + 100) / 13 = 44.5. The analysis of the task execution time quantiles shows that the stages 0 ÷ 2 have a long right tail of the distribution function, and the tasks of stages 4, 9, 10 have a long left tail. You can also notice that "garbage collection" and writing to disk before "Shuffle Write" does not have a significant impact on the execution time of the request.

CBF Stage	SF=500, N=40m: Amount(GB)/Number		
	Input	**Shuffle Read**	**Shuffle Write**
0	26,2/3 000 028 242		23,7/1 617 261 771
1	0,177/75 000 000		0,070/14 997 607
2	5,70/750 000 000		4,40/364 396 568
3		4,50/379 394 175	1,00/72 858 348
4		24,7/1 690 120 119	
9		24,7/1 690 120 119	0,109/5 658 551
10		0,006/337 532	
Sum:	**32,1/3 825 028 242**	**53,9/3 759 971 945**	**29,3/2 075 172 845**

Figure 8. Data volumes and the number of reads from the HDFS

Figure 8 shows the data volumes and the number of reads from the HDFS (Input) file system read from the local disks and transmitted over the virtual network during the "shuffle" stage when the Reduce (Shuffle Read) tasks written to local disks before the "Shuffle" by all the tasks of the corresponding stages. The arrows indicate the components of the corresponding volumes of Shuffle Read. It should be noted the high compression ratio of the original tables. On the average, it amounted to (see Tables 1 and 3) 500 * (26 + 149 + 641) / (32.1 * 1024) = 12.4.

Now consider the results of the same query Q3, but using CBF developed method (see Table 2, column 2). Fig. 9 shows the step (stage) CBF time and their implementation in Spark for SF = 500, the number of elements in the filter Bloom N = 40 million or probability of a false positive response filter equal to P = 0.01 [11]. Stage2 run on the task map, which scan set RDD "customer" (in MemoryTableScan RDD), has created Stage0 stage. Thus some data read from HDFS (0.079Gb / 1501 see Fig. 8 – … Is not C_ table). These tasks are placed recording RDD "customer" on the local drive (Shuffle Write) for further "shuffling" and compounds with a table at O_ Stage3 step. Tasks Stage1 and Stage2 are performed in parallel. Similar actions Spark performs on stage with Stage6 C1O1 table. So tasks Stage6 and Stage7 are performed in parallel too. Query execution time was 11 min, which is a 25/11 = 2.3 times smaller than the same query without using the Bloom filter (see Fig. 6).

	CBF	Time (min.) execution, SF=500, N=40 ml.
0	Reading C_ and Blom filter bloom1 (Stage0) creating	0,4
1	Reading O_ and bloom1 (Stage1) using	1,5
2	map C_ – join with O_ (Stage2)	0,3
3	C_▶◀O_ and bloom2 (Stage3) creating	0,6
6	map C_▶◀O_ –join with L_ (Stage6)	0,1
7	Reading L_ and bloom2 (Stage7) using	7,3
8	L_▶◀(C_▶◀O_), group by, order by (Stage8)	1,2
		SUM: 0,4+1,5+0,6+7,3+1,2=11 min

Figure 9. CBF Stages

Figure 10 shows the characteristics of the CBF tasks by performing steps (SF = 500, N = 40 million, P = 0.01). The arrow indicates that the stages 1 and 2 are performed in parallel on 13 executors. The average number of tasks performed by a single core, in this case is (100 + 24) / 13 = 9.5. Stages 6 and 7 are also performed in parallel.

SF=500 N=40	Quantity			Quantile task time , sec.					Avg. time per Task	
CBF Stage	nodes	executors	Tasks	min	0,25	0,5	0,75	max	Garbage Collect.	Shuffle Write
0	7	13	24	0,8	2	4	6	9	0,3	
1	7	13	100	3	4	9	12	19	0,1	0,063
2	1	2	24	0,004	0,029	0,041	0,1	3	0	0,3
3	7	13	14	16	19	19	19	20	0,8	
6	7	13	14	2	2	2	2	3	0,2	0,2
7	7	13	455	0,1	7	10	13	53	0,1	0,03
8	7	13	14	21	53	55	59	60	0,8	

Figure 10. Characteristics of the CBF tasks

Figure 11 shows the amount of data and number of entries Input, Shuffle Read and Shuffle Write. It should be noted that the volumes Shuffle Read and Shuffle Write using the method CBF smaller compared with NBF respectively 10.5 and 5.7 times (see Table 4).

CBF	SF=500, N=40m: Amount(GB)/Number		
Stage	Input	Shuffle Read	Shuffle Write
0	0,177/75 000 000		
1	5,70/750 000 000		0,944/75 783 663
2	0,079/1 501		0,070/14 997 607
3		1,014/90 781 270	
6	0,847/7 291		1,006/72 858 348
7	26,2/3 000 028 242		3,1/203 379 624
8		4,1/276 237 972	
Sum:	**33,0/3 825 037 034**	**5,11/367 019 242**	**5,11/367 019 242**

Figure 11. Characteristics of the CBF tasks

Fig. 12 shows the influence of the parameter N (mill) Bloom filter (the number of elements in the bitmap) to the performance indicators Q3 CBF query.

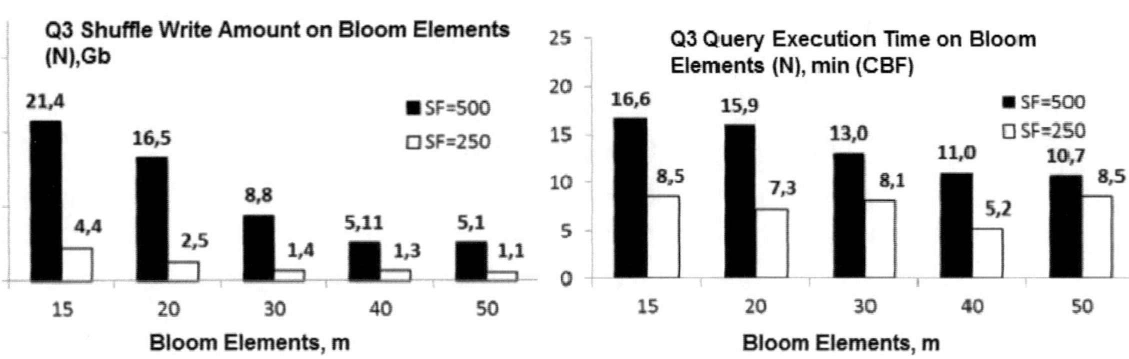

Figure 12. Shuffle write amount (a) and query execution time (b) on number of Bloom filter elements

Volume Shuffle Read (Shuffle Write) decreases substantially at increase N up to 30 million for SF=250 and up to 40 million for SF=500. Further growth N does not lead to a significant reduction of volume. This is due to the reduction in the number of false positives Bloom filter. Time shown on the fig. 8 b, is random (impact that cluster deployed in a virtual environment). But when SF = 500 is observed a tendency to reduce the query time until N = 40 Mill. In this experiment, the influence of N on the execution time of the request is lower than the amount of data on the "shuffling": for SF=500 at increase N from 15 million to 40 million volume decreased in 21.4/5.11= 4.2 times, and time – only in 16,6/11=1,5 times. This is due to the fact that the network between the nodes is a virtual, i.e. "Fast" (some components reside on the same physical server).

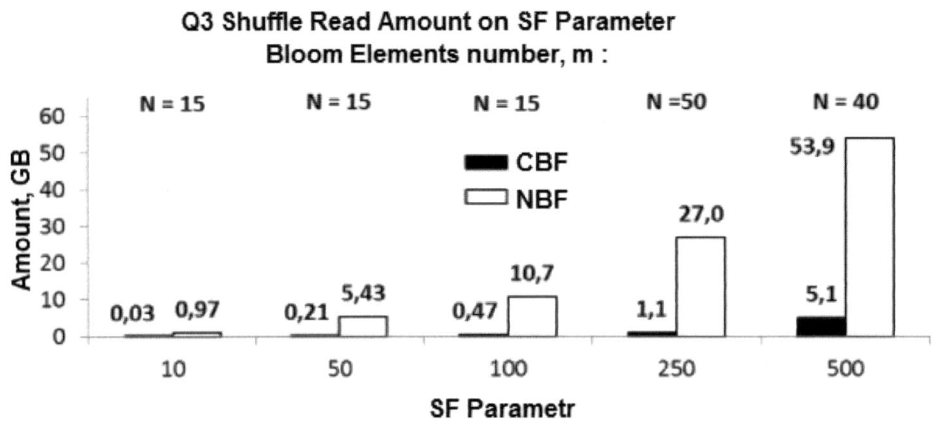

Figure 13. Shuffle Read Amount on SF

Figure 13 shows that the amount of intermediate data stored on the local disk and transmitted over the network in the process of "dicing" (Shuffle Read), is significantly higher for the method NBF than CBF.

Method CBF loses to method NBF on the time performances of the enquiry (fig. 10). Up to SF=250 NBF falls behind slightly CBF. But at SF=500 the advantage of the CBF is evident.

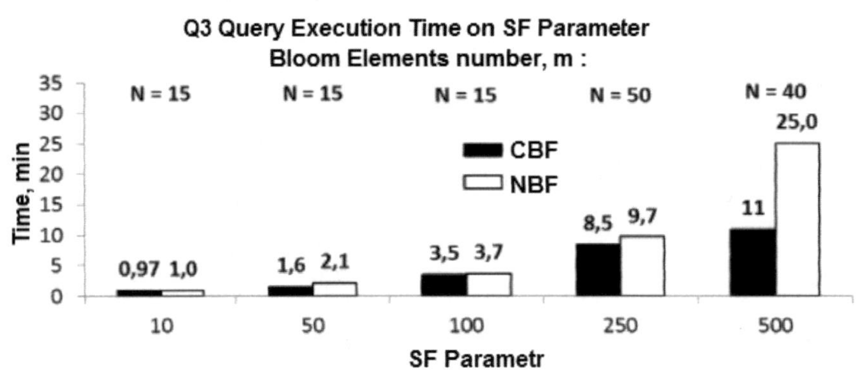

Figure 14. Query Execution time on SF

5. Conclusion

Spark drivers for the Q3 query of TPC-H test have been development both with and without Cascading Bloom Filter.

IOP Conf. Series: Journal of Physics: Conf. Series **913** (2017) 012011 doi:10.1088/1742-6596/913/1/012011

A series of full-scale experiments was carried out on a cluster of 7 work units implemented in the cloud architecture, in order to compare the methods of the CBF and NBF access to the data store by the example of Q3.

Using the developed CBF method, the amount of intermediate data stored on the disk and transmitted over the network when the query is executed is much smaller. So for SF = 500 these volumes differ by an order of magnitude (see Fig. 13). When executing a query, the load on the disk and the network for the CBF method is less. CBF wins by the time the query is executed. For SF = 500, the CBF method is 2.3 times better than the CBF over time (see Fig. 14). It is expected that the effect will be even higher if the cluster would not be implemented in a virtual environment. It is shown that the N parameter of the Bloom filter significantly affects the performance of the query using the CIFB method (see Fig. 12).

It is planned to develop a common cost model for the CBF access method and use the measurement results presented in this article to adapt the model and assess its adequacy.

References

[1] Venkat Goodwada, Dana Rao, Vijay Raghavan 2016 Renaissance DBMS: the problem of choosing *Open Systems*. DBMS **3** p 12–17

[2] Neljubin D and Kosjanenko G 2014 What's new in NEWSQL? Immersion in the latest databases *Hacker* **8** (**187**) p 127–131

[3] Dean J and Ghemawat S 2004 MapReduce: Simplified data processing on large clusters *The Sixth Conference on Operating System Design and Implementation, Berkeley, CA* p 10

[4] Riza S, Leserson Wu, Owen Sh and Wills D 2016 *Spark for Professionals: Modern Patterns for Large Data Processing* (St. Petersburg: Peter)

[5] Ermakov E Yu and Proletarskaya V A 2016 Method of access to the data store using MapReduce / Spark technology without caching of measurement tables in RAM *Information and measuring systems* **12** p 90–97

[6] Jaqueline Joice Brito, Thiago Mosqueiro, Ricardo Rodrigues Ciferri and Cristina Dutra de Aguiar Ciferri 2016 Faster cloud Star Joins with reduced disk spill and network communication *ICCS 2016. The International Conference on Computational Science* **80** p 74–85

[7] Grigoriev Yu A and Proletarskaya V A 2015 Method of early materialization of access to the repository using MapReduce technology *Informatics and control systems* **3** p 3–16

[8] Grigoriev Yu A and Proletarskaya V A 2016 Comparison of methods for processing requests to the repository using MapReduce technology *Informatics and control systems* **1** p 3–13

[9] Zhou G, Zhu Y and Wang G 2013 Cache Conscious Star-Join in MapReduce Environments. *Cloud-I '13 Proceedings of the 2nd International Workshop on Cloud Intelligence* p 1

[10] Bloom B H 1970 Space/time trade-offs in hash coding with allowable errors *Communications of the ACM* **13** p 422–426.

[11] Grigoriev Yu A, Proletarskaya V A and Ermakov E Yu 2017 Method of access to the data warehouse using SPARK technology with cascading Bloom filtering *Informatics and control systems* **1** p 3–14

[12] O'Neil P et al 2009 The star schema benchmark and augmented fact table indexing *Technology Conference on Performance Evaluation and Benchmarking* p 237–252

IOP Conf. Series: Journal of Physics: Conf. Series **913** (2017) 012012 doi:10.1088/1742-6596/913/1/012012

Models and algorithm of optimization launch and deployment of virtual network functions in the virtual data center

I P Bolodurina[1] and D I Parfenov[2]

[1] Department of Applied Mathematics, Orenburg State University, Pobedy avenue, 13, Orenburg, Russia
[2] Faculty of Distance Learning Technologies, Orenburg State University, Pobedy avenue, 13, Orenburg, Russia

e-mail: parfenovdi@mail.ru

Abstract. The goal of our investigation is optimization of network work in virtual data center. The advantage of modern infrastructure virtualization lies in the possibility to use software-defined networks. However, the existing optimization of algorithmic solutions does not take into account specific features working with multiple classes of virtual network functions. The current paper describes models characterizing the basic structures of object of virtual data center. They including: a level distribution model of software-defined infrastructure virtual data center, a generalized model of a virtual network function, a neural network model of the identification of virtual network functions. We also developed an efficient algorithm for the optimization technology of containerization of virtual network functions in virtual data center. We propose an efficient algorithm for placing virtual network functions. In our investigation we also generalize the well renowned heuristic and deterministic algorithms of Karmakar-Karp.

1. Introduction

The technology of cloud computing is built on the basis of the virtualization of the individual components that make up the data center infrastructure. Approaches to the organization of the virtualization layer are divided according to the levels of application of this technology. Typically, the following types of virtualization are distinguished: operating system, software, memory, data storage, databases and network. The most active development level is the virtualization of the network. It allows you to create a virtual environment to run multiple instances of user space within the same system used to run network service and functions within the cloud environment of virtual data center.

Nowadays virtual data center is representing as are a fairly complex multi-level mechanism that interacts in the course of its work with various objects and data flows of the network infrastructure [1, 2]. To analyze more data, it is necessary to apply the processing using methods used in Big Data.

One of the approaches to virtualization, which allows implementing this approach, is containerization technology. A container is an object that provides a user with access to necessary libraries and contains the required set of software for launching the development environment, a ready application or a service. Furthermore, the advantage of using virtualization on the basis of containers for placing applications and services in a network environment of a virtual data center is the lightness of the solution, which allows to reduce a number of costs and to increase the productivity of cloud-based service-oriented applications. As a containerization technology, the most effective is the control system based on Docker [3]. The advantage of this technology is its reliability, as well as the availability of open source code and a convenient API for sharing in a networked virtual data center environment. A significant drawback of containers is the inability to provide the proper level of isolation of data flows, as well as the lack of support for migration between compute nodes in real time.

A key difference between virtual machine virtualization and virtualization is the use of a hypervisor that emulates the hardware of a physical computing node. Within each virtual machine, a full-fledged operating system functions. All this leads to significant overhead in terms of resource consumption in a virtual data center.

As a result of investigations, we developed a solution that allows solving the problem of container migration, as well as contributing to optimization of overhead costs for computational resources for virtual data center infrastructures. The developed solution is based on hybrid virtualization, which is a combination of two approaches to the deployment virtual network functions in a software-managed infrastructure of a virtual data center: a container and based on virtual machines.

With resources virtualization technology development, the number of layers is steadily increasing, forming infrastructure decisions and being used in cloud computing technology. Nowadays, there can be defined up to six levels of software-defined infrastructure of virtual data centers, which use modern cloud platforms for deployment.

2. The level distribution model of software-defined infrastructure virtual data center

Let us introduce the level distribution model of software-defined infrastructure of virtual data center, which support the containerization method of network application and services disposal in cloud systems. The first level – the hardware component of any data center, which includes computational nodes (Nodes), file systems (Storages) and physical network units (NetObj). Let us introduce it as the set of decisions: $PhysLayer = \{Nodes, Storages, NetObj\}$.

The next level represents the software-defined layer. This layer consists of the same number of objects as the first level but the main difference is that all the infrastructure elements are dynamic, easily-transformed and adjusted within the limits of the physical database network environment. The second level can be presented as the next set of connections:

$$SDLayer=\{SDNodes, SDStorages, SDNetwork\}, \qquad (1)$$

where are: $SDNodes$ – software-defined computing units; $SDStorages$ – software-defined storages; $SDNetwork$ – software-defined network.

Above the layer of software-defined infrastructure, there is a level of the specific objects virtualization. Most of them are: computing nodes (VirtNodes), virtual data warehouses and also the elements of software-defined networks that are used in works of cloud platforms and consolidated in VirtNetwork multitude.

$$VirtLayer=\{VirtNodes, VirtStorages, VirtNetwork\}, \qquad (2)$$

In the software-defined infrastructure, computing nodes and data storages are more often presented as virtual machines that discharge the set of given functions.

To control such multi-layer infrastructure, the separate orchestration layer is needed (forth level). There are a number of functions in it and computing nodes and software systems are enabled for their execution. The main functions are orchestration of the virtualization objects (virtual machines and data storages) (ONodes, OStorages) and orchestration of the software-defined network (ONetwork), which include network function virtualization.

$$OrchLayer=\{ ONodes, OStorages, ONetwork\} \qquad (3)$$

IOP Conf. Series: Journal of Physics: Conf. Series **913** (2017) 012012 doi:10.1088/1742-6596/913/1/012012

The next level (Service level) represents the services used in the working process either the very cloud platform, or applications distributed there, for example, DBMS, Hadoop, Nginx and others.

$$ServiceLayer = \{Service_1, \ldots, Service_n\} \tag{4}$$

All the multitude of ServiceLayer network services working in the virtual data center infrastructure can be divided into two disjoint subsets $ServVM \cup ServDocker = ServiceLayer$. Services using virtualization on the base of other machines are in the first set (ServVM). In the second set (ServDocker), there are services realized on the base of containers under Docker control.

On the top level, there are network applications that are exploited by users for flexible scalability providing (AppLayer).

$$AppLayer = \{App_1, \ldots, App_m\} \tag{5}$$

As on the previous level, network applications Appi can be placed in containers, forming the AppVM set or use containerization forming the AppDocker set. At the same time $AppVM \cup AppDocker = AppLayer$.

Thus, the set of object of software-defined infrastructure can be divided into two groups by the methods of placing. Virtual objects using container placing method can be referred to the first group. Let us describe them in the following way:

$$Docker = \{ServDocker, AppDocker\}. \tag{6}$$

In the second group there are services and applications using virtual machines as a placing platform:

$$VM = \{ServVM, AppVM\}. \tag{7}$$

Before we talk about the ways of placing network applications and services, which realized virtual network functions in a virtual data center, we need to determine their structure, basic parameters, and key characteristics of their operation affecting the efficiency of their use. For this purpose, we have developed a generalized model of a virtual network function.

3. The generalized model of a virtual network function

The specific feature of virtual network function is the approach, where users have access to them and to their services; however, they do not know anything about their actual location. In most cases, users only know the address of the aggregation node and the application name. The orchestrator of virtual data center automatically selects the optimal virtual machine for the request, on which it is to be processed.

The generalized model of a virtual network function is a multilayer structure, described in a form of graphs to characterize the connections of individual elements. The model can be represented in the form of three basic layers, detailing the connections of the specific objects of virtual data center infrastructure: network applications, network services and network resources.

The network applications are a weighted directed acyclic graph of data dependencies:

$$NFVAppl = (G, V). \tag{8}$$

Its vertices G are tasks that get information from the sources and process it in accordance with the user requests; its directed edges V between corresponding vertices are a link between tasks in a schedule plan. The schedule plan is defined as a procedure which is prepared to follow the user's request (*SchemeTask*). Each vertex $g \in G$ is characterized by the following tuple:

$$NFVAppl = (Res, NAppl, Utime, SchemTask). \tag{9}$$

where are *Res* are the resource requirements; *NAppl* is the number of application instances; *Utime* is the estimated time for of the users' request execution; *SchemeTask* is a communication scheme of data transmission between sources and computing nodes.

Each directed edge $v \in V$ connects the network application with the required data source. It is characterized by the following tuple:

$$v=(u,v,Tdata,Mdata,Vdata,Qdata) \quad (10)$$

where u and v are linked vertices; *Tdata* is the type of transmitted data; *Mdata* is the access method to the data source; *Vdata* is the traffic volume estimated by the accessed data (in Mb), *Qdata* is the requirements for the QoS (quality of services).

The model is original, because it enables to calculate the consolidated assessment of its work with data sources for each network application. It allows to predict the performance of the network in whole virtual data center.

As mentioned above, a network service is one of the key slices within the generalized model of a virtual network function. The network service is an autonomous application. Generally, the network service is highly specialized and designed to perform a limited set of network functions. The advantage of using the network service is isolated data flow processing. The usage of network services as virtual network functions reduces the execution time for user requests. The network service is described as a directed graph of data dependencies. The difference lies in the fact that from the user's point of view, the network service is a closed system. The network service can be formalized as a tuple:

$$NetServ=(AgrIP,NameServ) \quad (11)$$

where are *AgrIP* is the address of aggregation computing node; *NameServ* is the network service name.

The aggregator of a network service selects an optimal virtual machine; it is executed on this machine. In addition, all its applications are distributed between predefined virtual machines or physical servers. In addition, all its network applications are distributed between predefined virtual machines or physical servers. Their new instances are scaled dynamically depending on the volume of incoming traffic.

In order to describe the placement of network applications and network services in the virtual data center infrastructure, we have also implemented the model of a network resource. A network resource is an object of a virtual data center, which describes the behavior and characteristics of the virtual network function, depending on its current state and parameters. The objects of virtual data center are disk arrays including detached storage devices, virtual machines, software-defined storages. In addition, each network service or network application imposes requirements on the number of computing cores, RAM and disk sizes, and the presence of special libraries on physical or virtual nodes used to launch their executing environments. Each network resource can be formalized as follows tuple:

$$Net\,\mathrm{Re}s=(T\,\mathrm{Re}s,Param,State,Core,Rmem,Hmem,Lib), \quad (12)$$

where are *TRes* is the type of resource; *Param* is the set of parameters; *State* is the state of resource; *Core* is the number of computing cores; *Rmem* is the size of RAM; *Hmem* is the size of a disk; *Lib* are libraries requirements.

The distinctive feature of the model suggested implies analyzing network resources from the user viewpoint and from the viewpoint of a software-defined infrastructure of the virtual data center. The model is innovative, since it simultaneously describes the virtual network function placements and the state of the virtual environment, taking into account the network topology.

4. The neural network model of the identification of virtual network functions in the infrastructure of the virtual data centers

We have chosen Kokhonen's network as a neural structure for modeling, since it is the most efficient in the clustering and classification of objects. Another important factor is the visualization of results; it enables to improve the understanding of the structure and character of data at early stages and to specify a neural network model further. Due to the peculiarities of network function virtualization, the support of classification in Kokhonen's network can be used to identify the uniform elements in network for further optimization of their placement. Kokhonen's network is trained by a method of consecutive approximations. Starting from the initial placement of objects selected randomly, the algorithm gradually improves it to supply the data clustering. Another advantage of Kokhonen's network is the opportunity to identify new clusters. The trained network detects clusters in the training data and refers all the data to certain clusters. If the network meets a set of data, which differ from any known samples, it will independently reveal a new cluster of elements then. This feature is very relevant, since it allows entering new network functions into the architecture of virtual data center without the actual change of algorithms of their distribution on physical and virtual computing nodes.

The principle of creating a neural network system to optimize the placement of network function virtualization in the multi cloud environment of the virtual data center is as follows. We have selected a number of criteria using the data obtained from the systems of virtual data center monitoring. This enables to both identify the virtual network function and assess its load on computing nodes. Criteria are formulated so that the answer could be always represented in the binary form. i.e. 1 is "Yes" or 0 is "No". The obtained data enable to form the vector of signals $E = \{e_1, e_2, \ldots, e_n\}$, which is placed at the entrance of a neural network. The vector of output values is similar, it has binary components.

The neural network is a two-dimensional matrix of neurons of dimension n (the number of inputs of each neuron) per m (the number of neurons). The number of inputs of each neuron is determined with respect to the number of criteria established earlier. The amount of neurons m coincides with the required number of classes and corresponds to the number of the unique network functions used in work of a multi cloud platform. The importance of each of the entrances to neuron is characterized by the numerical size called by weight. It is set in the form of matrix:

$$X = \begin{pmatrix} x^{11} & x^{12} & \ldots & x^{1j} \\ x^{21} & x^{22} & \ldots & x^{1j} \\ \ldots & \ldots & \ldots & \ldots \\ x^{i1} & x^{i1} & \ldots & x^{ij} \end{pmatrix}, \tag{13}$$

With the vectors of weight coefficients of connections $x^{ij} = \{x_1^{ij}, x_2^{ij}, \ldots, x_n^{ij}\}$ as its elements.

Kokhonen's network consists of three layers of neurons. The basis of the network is a covert Kokhonen's layer. However, in this research, we have offered a changed scheme of output neurons of Kokhonen's network to obtain the results to identify destination and, simultaneously, to find critical loading on a calculating node.

We offered to divide the covert layer of Kokhonen's neural network into two sets. The first set of neurons [1…K] is responsible for the identification of a network function placed in the virtual data center. The work of neural network changes input scales at the exit layer activates the linear function Y_1, which takes the value [0…n], where 0 means that the network object under study has no signs of a virtual network function, and values from 1 to n correspond to a particular network function identified by a neural network model. The second set of neurons [L…Z] analyzes the loading of the network object under study and initiates the function Y_2 at the exit, which take the values [-1,0,1], where 0 is a normal state, -1 means that the network function is idle or does not perform its functions, and 1 means that the network function is overloaded.

The basic criteria used as input data to detect virtual network functions are network records and events, data of the time of packages going through a network object, time of packet input and output, memory loading, the use of CPU, the intensity of dataflow, TTL and others. The data collected in the network are placed at the entrance of a neural network and create a full neural network. That being the case, we should simplify and verbalize a neural network by excluding some elements without the significant reduction of the detection quality.

To train a neural network, we have used the data obtained from the system of monitoring of virtual data center of the Orenburg State University. It includes 4 OpenFlow switches (2 x HP 3500yl, 2 x Netgear GSM7200), 8 computing nodes (32Gb RAM, 4 cores), 1 server (32Gb RAM, 8 cores) with OpenFlow controller and 1 server (32Gb RAM, 4 cores) for monitoring function. Routers connected compounds having the speed 1000 Mbit / s, and the computers are connected to a third level router via the second level network connections at 100 Mbit/s. We have chosen the most popular virtual network functions for the experimental study (Router, NAT, Firewall, Proxy, Switch, DPI).

Table 1. The result of experimental identification of virtual network functions

Virtual network function	The number of instances	Correct detection
vRouter	20	19 (98%)
vNAT	15	13 (94%)
vFirewall	18	17 (93%)
vProxy	25	24 (96%)
vSwitch	30	28 (93%)
vDPI	16	14 (87,5%)

The obtained data enable to conclude that the developed neural system had some difficulties with the detection of some network functions because of their small differences in the parameters chosen. This defect can be eliminated by the introduction of additional criteria to the initial model of neural network. Thus, the application of the developed neural network system enables to identify virtual network functions correctly in middle in 94% of cases, while the resource intensity is insignificant. This increases the efficiency of optimization for launch them in the infrastructure of a multi cloud platform.

5. The algorithm of optimization launch and deployment of virtual network functions in the virtual data center

The presented models allow us to choose the most suitable methods of launch and deployment instance of virtual network functions in virtual data center infrastructures based on current load and the incoming flow of requests. The main task of the distribution of virtual network functions is to choose the number of instances in time interval, which formulate as making a plan. When accessing the virtual network functions, the preparation of a plan is especially important. Due to the load, the compute nodes may vary widely over relatively short time intervals, and depends on the method of placement of the virtual data center infrastructure. To solve the optimization problem, an algorithm was developed, which monitors the virtual data center infrastructure, scheduling and launching applications and services. It is based on a biased random-key genetic algorithm (BRKGA). But different to BRKGA, the algorithm uses heuristic analysis queries flows and their classification depending on the application hosted in the virtual data center.

Our model of identification of virtual network functions allows optimizing their launch and deployment in the virtual data center. We will optimize the launch and deployment of the network functions found in the virtual data center by using Kokhonen's network by the following criteria: the current load created on computing nodes; resource intensity of a network function; the number of flows going through computing nodes. The main objective of virtual network functions launch and deployment is to choose the optimum number of the nodes to implement required functionality

IOP Conf. Series: Journal of Physics: Conf. Series **913** (2017) 012012 doi:10.1088/1742-6596/913/1/012012

as a software solution. Thus, there is a problem of resource planning. Planning is of particular relevance in the organization of dynamic topology in the virtual data center, since the load on computing nodes can change over a wide range at rather short intervals of time and depends on the chosen type of specific network functions placement. To solve the task of optimization, we have developed the algorithm for monitoring the infrastructure of the virtual data center as well as for the placement and launch of network functions. In comparison with the available analogs, the algorithm uses the heuristic analysis of traffic flows and their classification depending on the type of a network function.

The general algorithm has the following sequence of steps.

Step 1. To identify the arrangement of virtual network functions taking into account the topology of the network infrastructure of the virtual data center.

Step 2. To estimate the number of the launched copies of each virtual network function and to range them according to the popularity of the network infrastructure. The popularity is estimated towards traffic flows, which run through the launched copies of a virtual network function.

Step 3. To define the load on physical and virtual computing nodes created by each copy of function.

Step 4. To compare data and to define virtual network functions, which demand scaling or folding, using the data obtained in steps 1 and 2.

Step 5. To reconfigure the topology on the controller of the software-defined network, stop and release the occupied resources of virtual data centers for network functions, which require folding.

Step 6. To evaluate a method of placement for virtual network functions, which require scaling and creation of the maximum load on the infrastructure. To distribute the most loaded network function using a hybrid way of placement (the containers developed in the virtual machine). To transfer network functions, which are less loaded but require scaling, to the operating mode of a virtual machine.

Step 7. To provide the migration of the virtual machines with network functions on the least loaded computing nodes.

The approach in this algorithm for controlling the launch and deployment of virtual network functions enables to take into account a method of placement and to organize the work of the virtual data center with account for circulating traffic flows and regulate the number of the launched copies of each function.

6. Experimental part

The aim of the experimental research is to define the effectiveness of the use algorithm of placement the virtual network functions in infrastructure of the virtual data center. For each type of virtual network functions created images templates of container and virtual machine for deployment in infrastructure of the virtual data center. On the basis of statistical data using queries created by the flows generator are simulated requests from users to cloud applications. Flows of different intensity were used to evaluate the performance. In the first case, flows create a minimum load capacity (to evaluate response times and delays, which makes data center infrastructure) (exp 1). In the second case, workload was created. We generated the traditionally values of traffic for each cloud application (exp 2). In the third case, a developed algorithm of optimization launch and deployment of virtual network functions (exp 3).

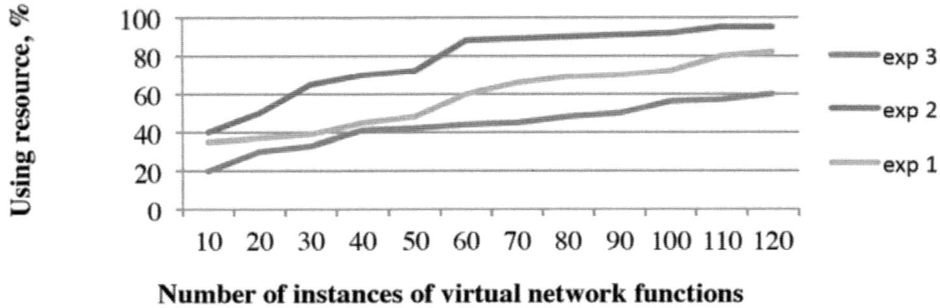

Figure 1. Computational experiment result

The investigation has shown that using static placement of virtual network functions in the containers on physical nodes is not effective, because it does not allow redistributing the load quickly. In addition, the movement of the container to another computing unit leads to a loss of current connections. Placing the virtual network functions based on virtual machines due to the flexibility of load balancing showed better results. However, loading on compute nodes has increased considerably due to the additional overhead associated with the use of virtual machines. The most effective placement of the study was the use of containers inside the virtual machines. This solution was possible to increase the density of placement virtual network functions within virtual data centers, as well as to allow placing containers, data services and network applications in close proximity to each other, thereby reducing the response time of applications on users' queries and thus increasing the efficiency of the cloud system.

7. Discussion

The variety of physical network devices of various vendors increases both capital expenditure and operational costs for the maintenance of data centers. The technology of network function virtualization allows solving this problem by the realization of network functions as software. The application of the NFV implies the use of the technology of network objects virtualization, which function as software and particular computing processes, or as complex infrastructures of cloud computing instead of hardware solutions.

The group of scientists headed by Liqiong Chen [4] analyzes the architecture and mechanisms of the interaction between the technology of network function virtualization and the software-defined networks. As noted in the research, if the number of the users who launch applications in the multi cloud network environment increases, there is a competition for resources. Besides, each user request is described by the relevant requirements to network environment from the viewpoint of productivity, safety, and the effective control of objects in virtual infrastructure.

Scientists from Arizona State University [5] have studied a multi cloud system. They offer an approach to the creation of network architecture based on the NFV technology as alternatives for traditional hardware network devices. However, this research does not solve the main problem concerning the methods of NFV placement on computing nodes.

Apart from a problem of the effective placement of network functions, the NFV technology has a number of disadvantages associated with the organization of the coordinated control of the entire network infrastructure of a virtual data center. To solve this problem, scientists from University of Wisconsin-Madison have proposed an approach, which is a framework of the OpenNF [6]. This approach provides the effective coordinated control of both the internal state of network functions and the state of data transmission network. However, this decision does not solve another important issue associated with the overall performance and load on the controller and objects of network functions.

IOP Conf. Series: Journal of Physics: Conf. Series **913** (2017) 012012 doi:10.1088/1742-6596/913/1/012012

The scientists from Nation Chung Cheng University have investigated the reduction of load on the controller to ensure the work of the NFV technology [7]. As a rule, in case of the software-defined network use, the controller classifies the traffic received from the ports on network nodes to define a path to network objects, which play a role of the virtual network functions. This process generates large volumes of traffic in the plane of control. The authors offer the expansion for the architecture based on the software-defined network to reduce the load created by network traffic on the controller due to the use of the NFV technology. The solution represents two-layer classification of traffic based on the OpenFlow protocol. Network events are analyzed in the plane of data instead of the plane of control.

The group of scientists headed by Hassan Jameel Asghar has developed a scalable system to work with the technology of network function virtualization in the multi cloud environment [8]. The authors have offered an abstract model of the standard network function distributed between several cloud platforms. The developed model is used as a basis for the SplitBox system. This system enables to increase the speed of package processing considerably in comparison with similar hardware solutions. An essential disadvantage of the concept is the resource intensity of this system. The overhead costs of the deployment and work of the Splitbox network functions demand the same quantity of computing resources as in case of hardware network devices. This disadvantage neutralizes the available advantages of the approach, because the problem of the effective placement of network functions in the multi cloud infrastructure of the virtual data center remains unsolved.

Thus, it is established that the technology of network function virtualization and the existing cloud solutions on based on the software-defined infrastructure of the virtual data center has a number of advantages, which enable to improve the quality of service in data transmission networks. However, today, there are adequate and effective solutions, that would enable to control the placement of the virtual network functions on physical and virtual computing nodes in the data center.

The analysis of scientific sources on the topic of the study has shown that:

a) so far, there are no effective algorithmic solutions for planning virtual machines and containers with virtual network functions, cloud services, application-oriented accounting topology of the computer system, and communication tasks schemes;

b) the existing solutions for managing virtual machines and containers on multi-cloud platforms plan computing tasks without subsequent adjustment of network to their communication schemes and use traditional routing methods;

c) the existing methods of data flow routing can be enhanced by taking into account the QoS requirements and distributed nature of a virtual network functions.

This demonstrates the novelty of the solutions offered by the project.

8. Conclusion

We propose an efficient algorithm for launch and deployment of virtual network functions in the infrastructure of a virtual data center. The problem of the optimization of launch and deployment of virtual network functions based on the template VM and containers with disabilities infrastructure virtual data center is reduced to the problem of packing in containers. We also generalize the well known/renowned heuristic and deterministic algorithms of Karmakar-Karp. We have developed an efficient algorithm to place VM by neural network optimization. Comparing the developed algorithm with the exact algorithm, we find that its approximate solutions do not differ much from the exact solutions. Thus, assessing the overall result of the algorithm performance gain can be obtained from 12 to 15% compared with conventional tools that are extremely effective at high intensities requests.

Acknowledgments

The research work was funded by Russian Foundation for Basic Research, according to the research projects No. 16-37-60086 mol_a_dk and 16-07-01004, and the President of the Russian Federation within the grant for state support of young Russian scientists (MK-1624.2017.9).

References

[1] Bolodurina I and Parfenov D 2017 Development and research of models of organization storages based on the software-defined infrastructure *Proc. 40th International Conference on Telecommunications and Signal Processing* (Barcelona: IEEE Press) p 1–5

[2] Bolodurina I and Parfenov D 2017 Methods and algorithms optimization of adaptive traffic control in the virtual data center *Proc. International Siberian Conference on Control and Communications, SIBCON 2017* (Astana: IEEE Press) p 1–5

[3] Bolodurina I and Parfenov D 2017 Development and Research of Models of Organization Distributed Cloud Computing Based on the Softwaredefined Infrastructure *Procedia Computer Science* **103** p 569–576

[4] Guisheng Fan, Huiqun Yu, Liqiong Chen, Dongmei Liu 2016 Modeling and analyzing cost and utilization based task scheduling for cloud application *Proc. 40th Annual Computer Software and Applications Conference (COMPSAC)* (Atlanta: IEEE Press) p 245–250

[5] Ruozhou Yu, Guoliang Xue, Vishnu Teja Kilari and Xiang Zhang 2015 Network function virtualization in the multi-tenant cloud *IEEE Network* (IEEE Press) p 42–47

[6] Aaron Gember-Jacobson, Raajay Viswanathan, Chaithan Prakash, Robert Grandl, Junaid Khalid, Sourav Das and Aditya Akella 2014 OpenNF: Enabling Innovation in Network Function Control *Proc. of the 2014 ACM conference on SIGCOMM* (Chicago: ACM) p 163–174

[7] Ying-Dar Lin, Po-Ching Lin, Chih-Hung Yeh, Yao-Chun Wang and Yuan-Cheng LaiAn 2015 An extended SDN architecture for network function virtualization with a case study on intrusion prevention *IEEE Network* (IEEE Press) p 48–53

[8] Hassan Jameel Asghar, Luca Melis, Cyril Soldani, Emiliano De Cristofaro, Mohamed Ali Kaafar and Laurent 2016 Mathy SplitBox: Toward Efficient Private Network Function Virtualization *Proc. of the 2016 workshop on Hot topics in Middleboxes and Network Function Virtualization* (Florianopolis: ACM) p 7–13

Institute of Physics
Dirac House, Temple Back
Bristol BS1 6BE UK

ISSN: 1742-6588
ISBN 978-1-5108-5040-8

CURRAN ASSOCIATES INC.
proceedings.com